RAPID REVISION NOTES

A LEVEL

PHYSICAL CHEMISTRY

RAPID REVISION NOTES

A LEVEL

PHYSICAL CHEMISTRY

BY S. LAKIN

*General Editor — Professor A.J.B. Robertson,
Professor of Chemistry, King's College,
London*

CELTIC
REVISION AIDS

Celtic Revision Aids
30-32, Gray's Inn Road,
London WC1X 8JL

First published 1982

ISBN 0-86305-117-0

Printed and bound in Great Britain by
Cox & Wyman Ltd, Reading

GENERAL EDITOR'S FOREWORD

The Rapid Revision 'A' Level Series is designed for students preparing for G.C.E. 'A' Level, Scottish Highers, Intermediate University and similar examinations. They are comprehensive and may be used either on their own for revision or as a complement to set books and text books. The notes are organised to help students remember facts and to pin-point areas of difficulty. Practice questions are included at the end of each section with answers to the numerical problems at the end of the text. Where necessary, clearly worked through examples have been included in the text.

I am sure students will find these books very helpful.

A J B Robertson

M.A., Ph.D., D.Sc., C. Chem., F.R.S.C.

Professor of Chemistry,
King's College London,
University of London

Formerly Fellow of St John's College
Cambridge

AUTHOR'S FOREWORD

This book deals specifically with the Physical Chemistry sections of the Advanced Level GCE Chemistry syllabus. It aims to aid the student's revision programme for the examination and is, therefore, a condensed form of the main points of the syllabus, rather than a full text book.

The notes which form the main body of each chapter, are organised into easily digested paragraphs, clearly subdivided by bold or italic headings. The notes are interspersed with example questions, which are worked through in the text to illustrate correct procedures. At the end of each chapter there is a list of practice questions, in the style of the main examining boards, for students to attempt on their own.

S Lakin

Contents

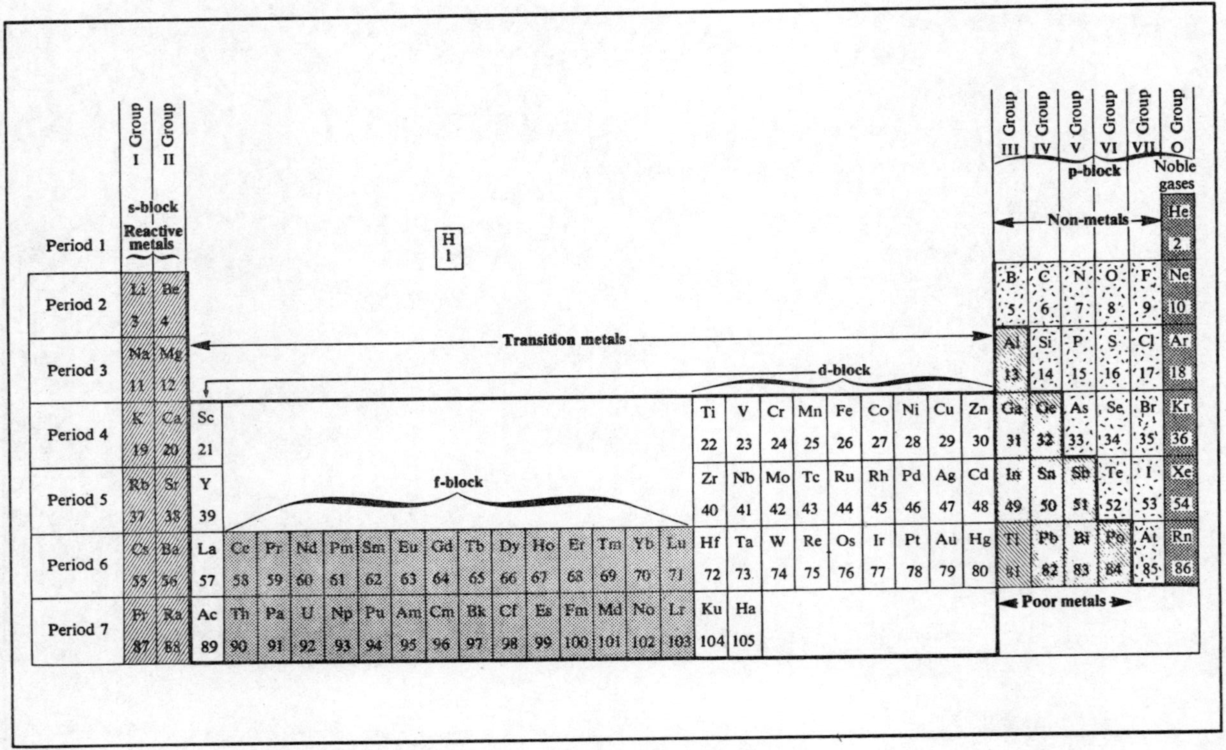
Group I
Group II
s-block
Reactive metals
Period 1
Period 2
Period 3
Period 4
Period 5
Period 6
Period 7
H 1
Li 3 Be 4
Na 11 Mg 12
K 19 Ca 20 Sc 21 Ti 22 V 23 Cr 24 Mn 25 Fe 26 Co 27 Ni 28 Cu 29 Zn 30 Ga 31 Ge 32 As 33 Se 34 Br 35 Kr 36
Rb 37 Sr 38 Y 39 Zr 40 Nb 41 Mo 42 Tc 43 Ru 44 Rh 45 Pd 46 Ag 47 Cd 48 In 49 Sn 50 Sb 51 Te 52 I 53 Xe 54
Cs 55 Ba 56 La 57 Hf 72 Ta 73 W 74 Re 75 Os 76 Ir 77 Pt 78 Au 79 Hg 80 Tl 81 Pb 82 Bi 83 Po 84 At 85 Rn 86
Fr 87 Ra 88 Ac 89 Ku 104 Ha 105
Ce 58 Pr 59 Nd 60 Pm 61 Sm 62 Eu 63 Gd 64 Tb 65 Dy 66 Ho 67 Er 68 Tm 69 Yb 70 Lu 71
Th 90 Pa 91 U 92 Np 93 Pu 94 Am 95 Cm 96 Bk 97 Cf 98 Es 99 Fm 100 Md 101 No 102 Lr 103
Transition metals
d-block
f-block
Group III
Group IV
Group V
Group VI
Group VII
Group O
p-block
Noble gases
Non-metals
He 2
B 5 C 6 N 7 O 8 F 9 Ne 10
Al 13 Si 14 P 15 S 16 Cl 17 Ar 18
Poor metals

1 Atomic Structure

Atoms are made up of three fundamental particles. Their relative masses and charges are given below. (The relative mass is the mass on a scale in which 1 atom of ^{12}C has a mass of 12 units.)

particle	relative mass	relative charge
proton	1	+1
neutron	1	0
electron	$\frac{1}{1840}$	−1

Many other subatomic particles such as positrons, mesons, anti-protons and neutrinos etc. have also been identified but their study is not relevant to the scope of this book.

The atom is electrically neutral so the number of protons equals the number of electrons. The **atomic number**, Z, of an atom is the number of protons in its nucleus and determines its identity and position in the Periodic Table.

The **mass number**, A, of an atom is the number of protons plus the number of neutrons in its nucleus.

The atomic number and mass number of an element X are written as $^{A}_{Z}X$, e.g. $^{35}_{17}Cl$.

A brief outline of the historical development of atomic theories

(a) **Dalton** – Dalton (1807) suggested that atoms were indivisible particles. His ideas had to be modified in the light of work on electrolysis (mainly by Davy and Faraday) and on the discharge of gases (Crookes and J J Thomson).

(b) **J J Thomson** – J J Thomson was the first to measure the charge to mass ratio of the particles making up cathode rays (1897) which Stoney later called electrons. At very low pressures gases will conduct a current if the applied potential is sufficiently high. Thomson observed that the zinc sulphide strip at the end of the tube fluoresced when struck by a stream of cathode rays – rays emanating from the cathode and accelerated towards the anode and through its slit. They were shown to be negatively charged by the direction of deflection when the electrical and magnetic fields were applied. By measuring the amount of deflection caused by a certain electrical field and the size

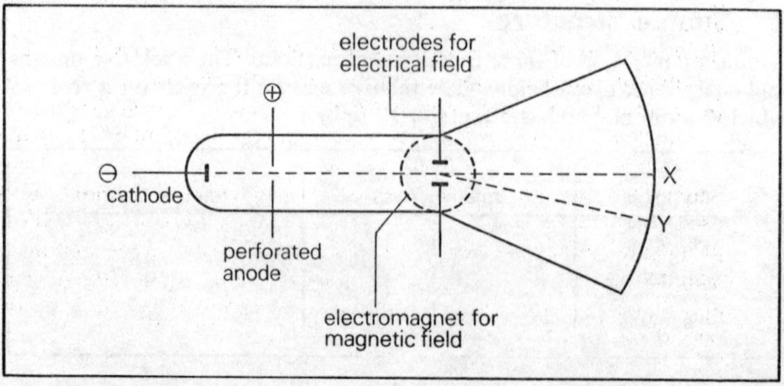

of the magnetic field required to return the ray to its original position, Thomson calculated the charge to mass ratio of the particles in the ray. The gas present in the tube is ionized as its external electrons are lost, e.g. He forms He^+ and He^{2+} ions. These ions pass to the cathode. The charge on the electron was found in 1909 by Millikan using the 'oil drop experiment'. From this result and his own value for the charge to mass ratio of the electron, Thomson concluded that the mass of the electron was $\frac{1}{1836}$ that of the hydrogen atom. He pictured the atom as a positive spongy material in which the small negative electrons were embedded – the 'plum-pudding' model.

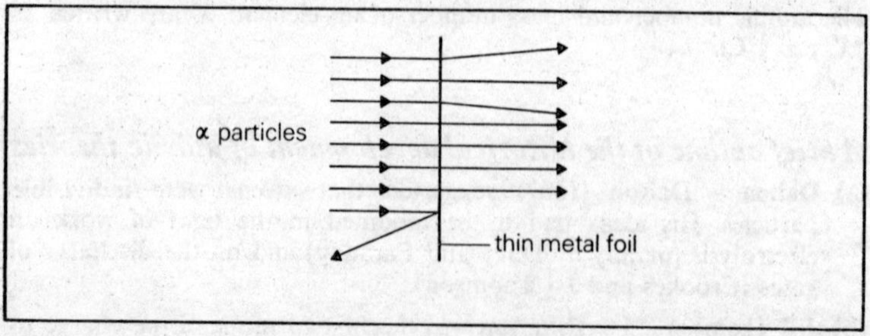

(c) In 1911, **Rutherford** proposed the existence of a positively-charged nucleus as a result of experiments by Geiger and Marsden (1909). They bombarded thin metal foils (platinum and gold) with a beam of fast moving, very small, positively-charged particles (α particles from a radioactive source). It was found that when observed as scintillations on a zinc sulphide screen the majority of the particles passed straight through the foil undeviated even though the foil was several hundred atoms thick. This indicated that the atom must be mainly empty space. A few α particles were slightly deflected and a very few (1 in 20000) were deflected right back through angles of greater than 90°. From these results Rutherford suggested that the atom had:

(i) A small positively-charged nucleus of diameter about 10^{-12} mm making up the major part of the mass of the atom.

(ii) A number of negatively-charged electrons revolving around the nucleus – the number of electrons equalling the number of positive charges which were called protons.

Rutherford's theory was found to be unsatisfactory as it could not be used to explain the non-integral values of many atomic masses without proposing the existence of electrons and protons with fractional mass. Van der Broek pointed out that the number of protons in the nucleus approximately equalled the numbered position of the atom in the periodic table (its atomic number Z).

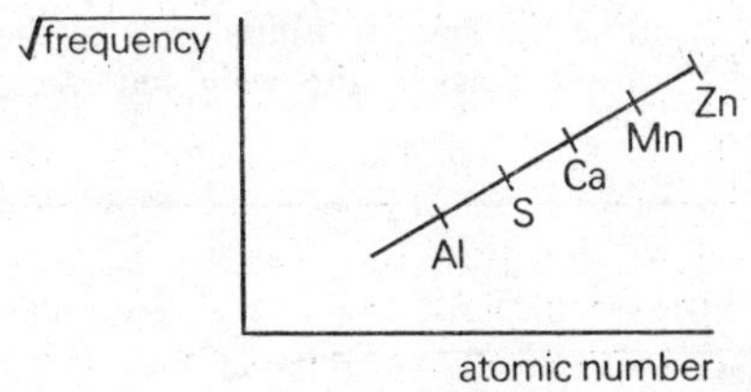

(d) **Moseley's** experiments of 1914 with X-rays gave fundamental significance to the term atomic number. He found that when a metallic element was bombarded with an electron beam, X-rays were emitted and that their frequency depended only on the element used. The frequency, v, was related to the atomic number Z of the element by the equation $\sqrt{v} = a(Z - b)$ where a and b are constants. This suggested that the atomic number of an element was related to some fundamental property of the atoms of that element. It was gradually accepted that the atomic number was, in fact, the number of positive charges or protons in the nucleus. Rutherford suggested the existence of neutrons after the discovery of isotopes in 1913.

(e) **Chadwick** – Experimental evidence for the neutron was not obtained until 1932 when Chadwick bombarded beryllium with α particles. He obtained neutrons as a highly penetrating radiation consisting of neutral particles each of mass almost identical to that of a proton.

Isotopes

Isotopes of an element are atoms which have the same atomic number but different mass number (i.e. the same number of protons and electrons but different numbers of neutrons). They have the same chemical properties but different physical properties such as boiling point, melting point, density etc.

Some of the isotopes of uranium are

$^{238}_{92}U$ 92p 146n 92e
$^{235}_{92}U$ 92p 143n 92e
$^{234}_{92}U$ 92p 142n 92e

The relative atomic mass A_r of the element is the weighted mean of the isotope masses on a scale in which 1 atom of carbon ($^{12}_{6}C$) has a mass of 12 units.

Determination of atomic masses
The atomic mass of an element is found by using a mass spectrometer. The principle is to determine the relative abundance of the isotopes of the element and their isotopic masses; the weighted mean of these is then the atomic mass.

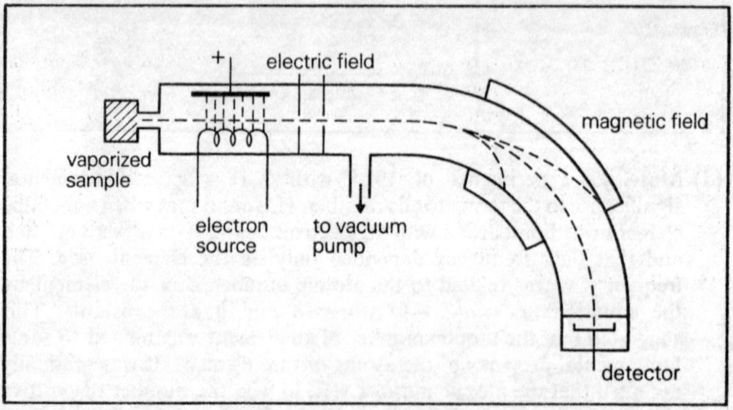

The element is vaporized and the gaseous atoms pass into the apparatus which is maintained under a high vacuum. As they pass between the anode and heated cathode of the ionization chamber, the atoms are bombarded by electrons freed from the cathode and travelling towards the anode. Ionization occurs.

$$X(g) + e\ (\text{fast}) \longrightarrow Z^+(g) + 2e\ (\text{slow})$$

The positive ion stream is accelerated by the electric field. For given electric and magnetic fields only ions of a particular mass/charge will reach the detector. By gradually increasing the strength of the magnetic field, ions of different masses may be brought successfully to the detector.

Their masses are calculated from the known applied fields, and their relative abundance from the relative magnitudes of the current produced in the detector. The ions are usually singly charged.

Mass spectrometer trace for naturally occurring lead.

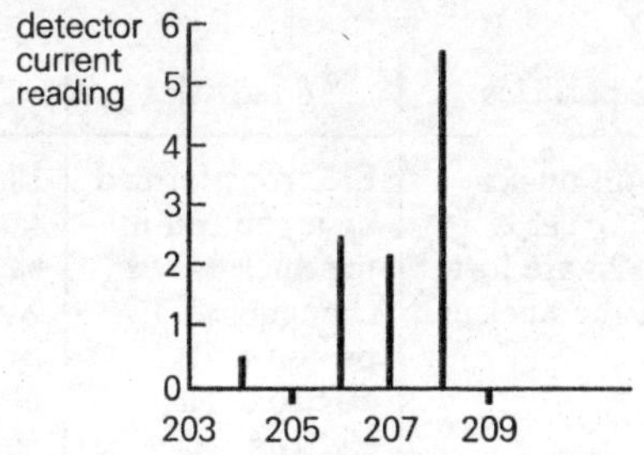

Radioactivity

Radioactivity is the disintegration of atomic nuclei with the emission of particles and/or rays. It was discovered in 1896 by Becquerel who found that compounds of certain elements emitted penetrating radiation which fogged photographic plates.

There are fifty naturally occurring radioactive isotopes belonging to 3 decay series – the uranium, thorium and actinium series. Stable nuclei lie within a stability belt as shown. Light stable isotopes have approximately equal numbers of protons and neutrons. Heavier stable isotopes have more neutrons than protons.

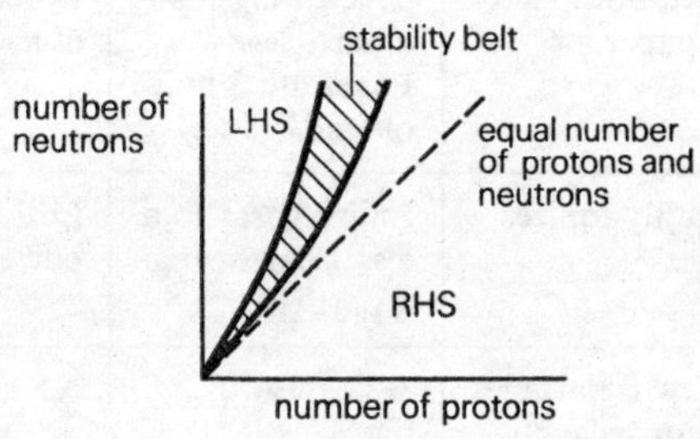

Isotopes on the right hand side of the stability belt must emit at least one α particle to attain stability i.e. to lie within the belt. (Combinations of α particles with β particles and γ rays may be given off.)

Isotopes on the left hand side of the stability belt attain stability by emission of at least 1 β particle.

Types of radioactive disintegration

The nature and properties of the 3 types of disintegration are summarized in the table.

	α particles	β particles	γ radiation
Nature	Helium nuclei (He^{2+}), i.e. $2n + 2p$ are lost from the nucleus.	Electrons formed by a neutron in the nucleus changing to a proton and electron. The electron is ejected.	Electromagnetic radiation of shorter λ than X-rays. Formed by a readjustment of the nucleus after α or β emission.
Charge	2+	1−	No charge
Effect of emission	An element 2 places back in the periodic table is formed. AN decreases by 2. MN decreases by 4.	An element 1 place on in the periodic table is formed. AN increases by 1. MN remains the same.	No new element is formed but a slight decrease in mass occurs. AN and MN remain the same.
Velocity and penetrating power	0·05 that of light. Little penetrating power. Stopped by paper.	0·3 to 0·99 that of light. More penetrating than α particles. Penetrate 3 mm of lead.	Velocity of light. Very penetrating. Penetrate 150 mm of lead.
Ionizing power	Readily ionizes gases.	Ionize gases to a lesser extent than α particles.	Little ionizing power.
Scattering by metal foils	A few α particles are deflected.	A beam of β particles is diverged. Repelled by the electrons on the metal atoms.	No effect.

	α particles	β particles	γ radiation
Electric and magnetic fields	Attracted to the negative plate and South Pole. The extent of deflection depends on the velocity.	Attracted to the positive plate and North Pole.	Not attracted as uncharged.
Examples	$^{226}_{88}Ra \rightarrow ^{222}_{86}Rn + ^{4}_{2}He$	$^{214}_{82}Pb \rightarrow \beta + ^{214}_{83}Bi$	$^{60}_{27}Co \rightarrow ^{60}_{27}Co + \gamma$

Characteristics of radioactive decay

1 The rate of decay is independent of temperature and pressure and of whether or not the atom is free or combined.

2 The rate of decay is proportional to the number of unstable nuclei present, i.e. it is a first order reaction.
The half life $t_{1/2}$ is the time taken for half of the radioactive atoms to decay. Values of $t_{1/2}$ can vary considerably. For example for uranium-238, $t_{1/2} = 4.4 \times 10^9$ years and for polonium-214, $t_{1/2} = 1{\cdot}5 \times 10^{-4}$ sec. The decay can be represented graphically.

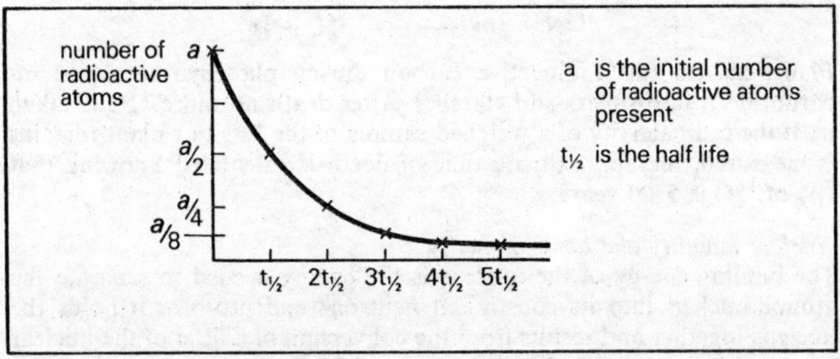

Detection of radioactivity

1 Cloud chambers (for detecting α and β rays).
The chamber consists of a vessel containing a gas supersaturated with water vapour. The α and β rays ionize the gas and droplets of water condense on the ions. The trace of the particle appears as a misty line.

2 Geiger-Müller counters (for detecting α, β and γ radiation).

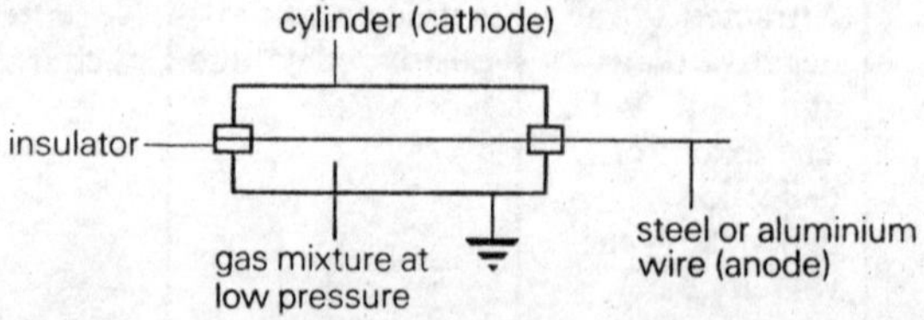

In the diagram above, the central wire has a potential of up to 1 kV with respect to the walls of the cylinder which are earthed.
The radiation ionizes the gas and enables a pulse of current to flow between the wire and cylinder. The pulse is amplified and recorded.

3 Scintillation counters (for detecting α, β and γ radiation).
When α and β particles and γ rays strike certain substances (phosphors) a flash of light is produced (fluorescence). This is converted to an amplified electric current and recorded. Typical phosphors are zinc sulphide (for α particles), anthracene (for β particles) and sodium iodide (for γ rays).

Radioactive carbon dating

$^{14}_{6}C$ is a radioactive isotope of carbon produced in the atmosphere when $^{14}_{7}N$ absorbs a neutron which has been activated by cosmic radiation. A proton is emitted.

$$^{14}_{7}N + ^{1}_{0}n \longrightarrow ^{14}_{6}C + ^{1}_{1}p$$

Plants absorb this radioactive carbon during photosynthesis and incorporate it into sugars and starches. After death no more $^{14}_{6}C$ is taken in. If the radioactivity of a weighed sample of the fossil or plant remains is measured, the approximate time of death is calculated knowing that $t_{1/2}$ of $^{14}_{6}C$ is 5700 years.

Nuclear stability and binding energy

The binding energy of the nucleus is the energy needed to separate the atomic nucleus into its constituent neutrons and protons. It holds the nucleus together and results from the conversion of a little of the nuclear mass into energy. Einstein's equation for the interconversion of matter and energy applies.

$$E = mc^2$$

where E is the energy in J, m is the mass in kg and c is the velocity of light, i.e. 3×10^8 m s^{-1}. For example consider 1 helium nucleus which consists of 2 neutrons and 2 protons.

Mass of helium nucleus (on $^{12}_{6}C = 12$ scale) = 4·0017 amu (atomic mass units).

Total mass of 2 neutrons and 2 protons

$$= (2 \times 1{\cdot}0073) + (2 \times 1{\cdot}0087) \text{ amu}$$
$$= 4{\cdot}0320 \text{ amu.}$$

$\therefore$ the loss in mass when $2n$ amd $2p$ join together to form a helium nucleus = 4·0320 − 4·0017 amu = 0·0303 amu.
This is known as the **mass defect**.
Loss in mass in forming 1 mole of He nuclei from neutrons and protons = 0·0303 × Avogadro's Number amu.

But $$1 \text{ amu} = \frac{1}{\text{Avogadro's Number}} \text{ g}$$

$\therefore$ $$\text{loss in mass} = 0{\cdot}0303 \text{ g}$$
$$= 0{\cdot}0303 \times 10^{-3} \text{ kg.}$$

Using $E = mc^2$.

$$\text{Energy released} = 0{\cdot}0303 \times 10^{-3} \times (3 \times 10^8)^2 \text{ J}$$
$$= 2{\cdot}73 \times 10^{12} \text{ J.}$$

This is the binding energy of 1 mole of helium nuclei.

Artificial transmutation of the elements
Radioactive isotopes of many elements can be made by bombarding the element or its compound with high energy particles such as protons, neutrons or α rays, e.g.

(i) $${}^{14}_{7}\text{N} + {}^{4}_{2}\text{He} \longrightarrow {}^{17}_{8}\text{O} + {}^{1}_{1}\text{H}$$

or $${}^{14}_{7}\text{N} \xrightarrow{(\alpha,\, p)} {}^{17}_{8}\text{O}$$

(ii) $${}^{27}_{13}\text{Al} + {}^{4}_{2}\text{He} \longrightarrow {}^{30}_{15}\text{P} + {}^{1}_{0}n$$

or $${}^{27}_{13}\text{Al} \xrightarrow{(\alpha,\, n)} {}^{30}_{15}\text{P}$$

Many artificial radioactive isotopes are produced in the atomic pile by bombardment of compounds contained in magnesium alloy cans with slow moving neutrons, e.g.

$${}^{27}_{13}\text{Al} + {}^{1}_{0}n \longrightarrow {}^{28}_{13}\text{Al} + \gamma$$

The ${}^{28}_{13}\text{Al}$ decays by β emission

$${}^{28}_{13}\text{Al} \longrightarrow {}^{28}_{14}\text{Si} + \beta$$

The method has been used to synthesize elements with atomic numbers greater than 92 which do not occur naturally, e.g. Curium (Cm) is made by bombarding plutonium with α particles.

$${}^{239}_{94}\text{Pu} + {}^{4}_{2}\text{He} \longrightarrow {}^{242}_{96}\text{Cm} + {}^{1}_{0}n$$

(${}^{239}_{94}\text{Pu}$ is synthesized as shown ${}^{238}_{92}\text{U} + {}^{1}_{0}n \rightarrow {}^{239}_{94}\text{Pu} + 2\beta$).

Energy changes during nuclear reactions

If the isotopic masses of the particles involved in a particular nuclear reaction are known then the energy change which has occurred can be found, e.g.

$$^{1}_{1}\text{H} + ^{7}_{3}\text{Li} \longrightarrow 2\,^{4}_{2}\text{He}$$

Relative isotopic masses:

$$^{1}_{1}\text{H} = 1{\cdot}0007825; \quad ^{7}_{3}\text{Li} = 7{\cdot}01600; \quad ^{4}_{2}\text{He} = 4{\cdot}00250.$$

Decrease in mass = $(1{\cdot}007825 + 7{\cdot}01600) - (2 \times 4{\cdot}00260)$ amu.
Consider 1 mole of $^{1}_{1}$H and 1 mole of $^{7}_{3}$Li.
Decrease in mass = $0{\cdot}018625 \times 10^{-3}$ kg.

Using Einstein's equation $E = mc^2$,

$$\text{energy released} = 0{\cdot}018625 \times 10^{-3} \times (2{\cdot}998 \times 10^{8})^2 \text{ J}$$
$$= 1{\cdot}67 \times 10^{12} \text{ J (correct to 3 sig. figs.).}$$

This energy appears as the production of heat (the products move more quickly) and/or as the release of γ radiation.

Nuclear fission

During nuclear fission a heavy isotope is split by a neutron into lighter nuclei with the emission of 2 or more fast moving neutrons and energy (often in the form of γ radiation). The neutrons produced react with more atoms of the heavy isotope setting up a chain reaction. One of the more common processes used in nuclear power stations involves the fission of $^{235}_{92}$U. The products of fission may vary but an example of one reaction which may occur is given by the following equation.

$$^{235}_{92}\text{U} + ^{1}_{0}n \longrightarrow ^{95}_{42}\text{Mo} + ^{139}_{57}\text{La} + 2\,^{1}_{0}n + 7\beta$$

Natural uranium contains the isotopes ^{238}U and ^{235}U in the ratio 140 : 1. ^{238}U absorbs neutrons and so the ore is first enriched with ^{235}U. One method of separating the isotopes involves diffusion. The uranium is converted to the gaseous hexafluoride UF_6 and by Graham's law the lighter $^{235}UF_6$ diffuses more quickly.

The rods of enriched uranium are placed in magnesium alloy cans within a block of graphite. Graphite acts as a moderator slowing down the secondary neutrons from the ^{235}U fission. This prevents their absorption by ^{238}U which absorbs fast moving neutrons only. If the chain reaction occurs too quickly an explosion may result and so control rods of cadmium or boron (good neutron absorbers) are available for lowering into the pile.

The energy liberated in the form of heat is removed by a coolant such as carbon dioxide or liquid sodium. The heat is used to boil water for producing steam to drive turbines.

Nuclear fusion

When 2 small atoms are made to fuse together a large amount of energy is evolved. This process occurs in stars and in the hydrogen fusion bomb, e.g.

$$^{3}_{1}H + ^{2}_{1}H \longrightarrow ^{4}_{2}He + ^{1}_{0}n$$

Controlled fusion processes have not, as yet, been achieved as very high temperatures are required to initiate the process.

Uses of radioisotopes

1 Treatment of cancer – the radiation from radium, cobalt-60 and iodine-131 is used to kill cancer cells.

2 Tracer techniques – radioisotopes are used to detect leakages and blockages in underground pipes etc. and in elucidating reaction mechanisms in chemical and biological systems.

3 Monitoring processes – beta and gamma radiation is used in measuring and controlling the thickness of materials and levels of substances in containers.

4 Radiography – gamma radiation is often used instead of X-rays as the equipment is cheaper and more manoeuvrable.

Electronic configuration

Summary

1 Electrons exist in discrete energy levels only. Each major level is given the symbol n which is known as the **principal quantum number**. The energy associated with each level increases as n increases, the lowest energy level being the one nearest the nucleus.

(a) Electrons fill the lowest energy levels first and when an atom is ionized electrons of the highest energy (largest n value) are lost first.

(b) When electrons are promoted from lower to higher energy levels, energy is absorbed. This is released when they return to their original levels.

(c) The difference in energy between successive energy levels decreases as n increases, i.e. the energy levels in the atom get closer together and finally converge at a convergence limit.

2 Each principal energy level can hold up to a maximum number of electrons given by the formula $2n^2$ where n is the principal quantum number, i.e. the maximum numbers when $n = 1, 2, 3$ are 2, 8, 18 respectively.

3 When $n > 2$ not all the electrons with the same n value have the same energy as subsidiary energy levels exist. Each subsidiary level can hold a maximum of 2 electrons and these must have opposite spin. The subsidiary energy levels available for each *n* value are listed below.

principal quantum number	energy levels available	illustration of levels full of electrons	total
1	1s	⇅	2
2	2s	⇅	8
	3 2p	⇅ ⇅ ⇅	
(the 2p levels have equal energy – are degenerate but have slightly higher energy than the 2s level)			
3	3s	⇅	18
	3 3p	⇅ ⇅ ⇅	
	5 3d	⇅ ⇅ ⇅ ⇅ ⇅	
	(the 5 3d levels are degenerate)		

The 3*d* levels have a slightly higher energy than the 3*p* than the 3*s*. However, the 3*d* levels have slightly higher energy than the 4*s* level so the 4*s* is filled first. The relative energies of the levels are shown diagrammatically below.

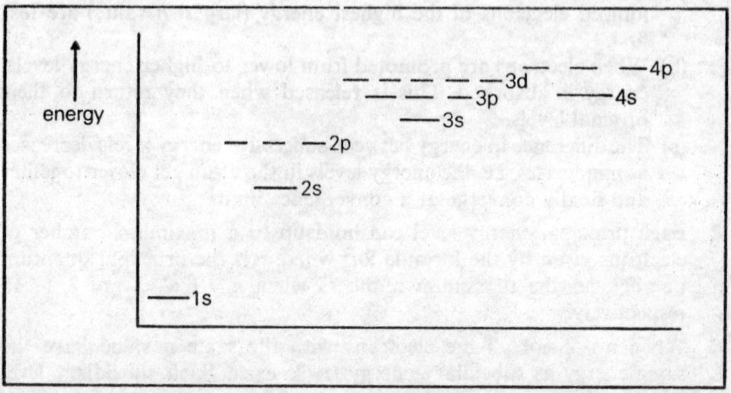

4 When energy levels are filled or 'built-up' with electrons, 3 principles are followed.

(a) Lowest energy levels are filled first, e.g. the electronic configuration of an atom of AN 23 is $1s^2 2s^2 2p^6 3s^2 3p^6 4s^2 3d^3$.

(b) The Pauli exclusion principle operates. This states that no 2 electrons can occupy the same subsidiary energy level unless they have opposite spin.

(c) Hund's rule of maximum multiplicity applies. This states that when electrons enter *p* or *d* levels they do so singly until each contains 1 electron. Pairing up then occurs, e.g. 5 electrons distributed in the 3*d* levels will have the configuration

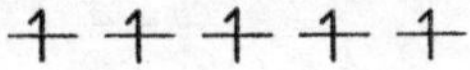

not

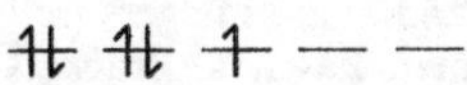

Evidence

1 Atomic spectra

The atomic spectrum of the simplest atom, hydrogen, is considered. When an electric discharge is passed through hydrogen at low pressure, a pink glow is observed. When this is examined through a spectroscope a few sharp coloured lines or a **line emission spectrum** is obtained.

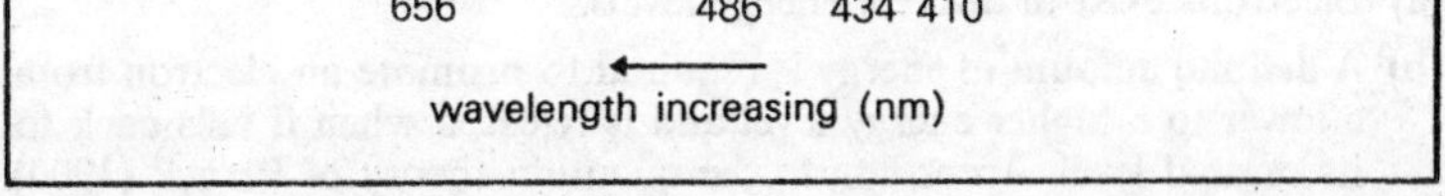

Similar series of lines are obtained in other regions of the spectrum (special detecting instruments are used).

Series of lines	**Region observed**
Lyman	Ultraviolet
Balmer	Visible
Paschen	Infrared
Brackett	Far infrared
Pfund	Far infrared

Features of the spectrum

(a) It consists of discrete lines of definite wavelength or frequency. For all electromagnetic radiation

$$\text{frequency} = \frac{\text{velocity}}{\text{wavelength}}$$

where the velocity = $2{\cdot}998 \times 10^8$ m s^{-1}.

(b) The lines get closer together at higher frequency and eventually coalesce to a convergence limit.

(c) Each series of lines is found to obey the equation

$$\frac{1}{\lambda} = R\left(\frac{1}{n^2} - \frac{1}{m^2}\right)$$

where λ is the wavelength in m; R is the Rydberg constant ($1{\cdot}09 \times 10^7$ m^{-1}); n and m are integers (m is always greater than n). For the Lyman, Balmer, Paschen, Brackett and Pfund series, n has the value of 1, 2, 3, 4 and 5 respectively. Thus for the Balmer series, when $1/\lambda$ is plotted against $1/m^2$ where $m = 2 + 1, 2 + 2, 2 + 3$, for the α, β, γ lines respectively, a straight line graph is obtained.

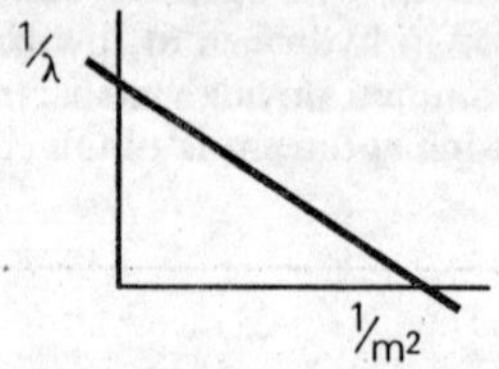

since $1/\lambda = R\left(1/2^2 - 1/m^2\right)$
the gradient = $-R$

Interpretation of the spectrum

To explain the spectrum the following principles must pertain.

(a) Electrons exist in discrete energy levels.

(b) A definite amount of energy is required to promote an electron from a lower to a higher energy level and is released when it falls back to its orginal level. According to the quantum theory of Planck (1900) this energy is emitted in discrete pockets or 'quanta'. Each quantum of radiation has an associated frequency given by the equation

$$E = h\nu$$

where E is the energy in kJ mol^{-1}; ν is frequency in s^{-1}; h is Planck's constant $3{\cdot}99 \times 10^{-13}$ kJ s mol^{-1}. Consider an electron falling from the $n = 4$ to $n = 2$ levels which have associated energies of E_4 and E_2 respectively. Then the amount of energy released, $\Delta E = E_4 - E_2$. Using Planck's theory a line of definite frequency is obtained in the emission spectrum for this particular transition.

The diagram shows the electron transitions associated with lines in the Lyman, Balmer and Paschen series.

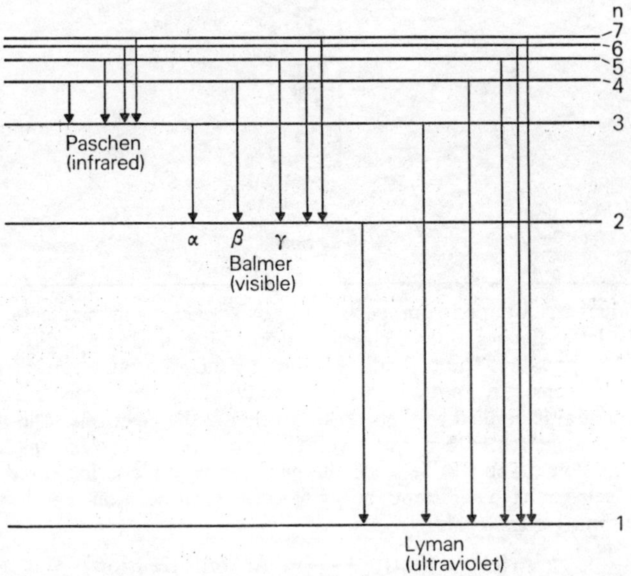

the Greek letters in the Balmer series refer to the lines in the visible spectrum

(c) The energy levels in the atom converge and finally coalesce at $n = \alpha$, so the lines in the spectrum converge to a convergence limit at the higher frequency end of the spectrum. If an electron is promoted from its ground state to beyond $n = \alpha$, then ionization occurs. The ionization energy of the hydrogen atom is calculated from the frequency of the convergence limit in the Lyman series by using the relationship $E = \hbar\nu$.

2 Ionization energy *IE*

The *IE* of an atom or ion is the energy required to remove 1 electron from the atom or ion in its gaseous state. (Values are quoted for 1 mole of the sample), e.g.

$$Na(g) \longrightarrow Na^+(g) + e \qquad \Delta H = 500\ kJ\ mol^{-1}$$
$$= \text{1st } IE \text{ of Na}$$

$$Na^+(g) \longrightarrow Na^{2+}(g) + e \qquad \Delta H = 4600\ kJ\ mol^{-1}$$
$$= \text{2nd } IE \text{ of Na}$$

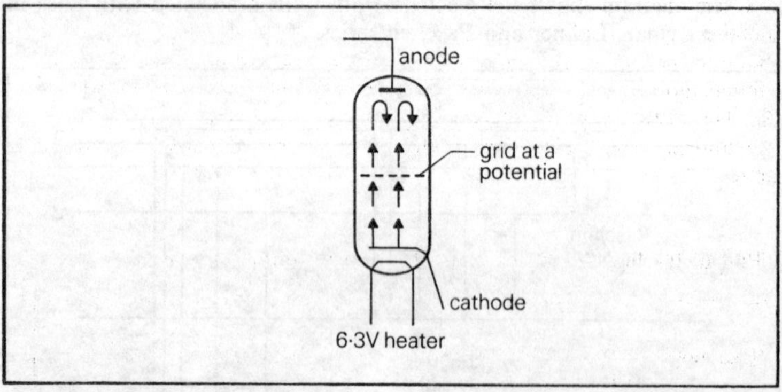

IE values are obtained from spectra or by using electron impact techniques. In the latter the gas of the element concerned is placed in a radiovalve at very low pressure. The cathode is heated by means of an electric current and electrons are emitted into the body of the valve as a stream. A potential is applied to the grid to accelerate the electrons. The anode is made negative with a small voltage and repels the electrons so no current flows. The PD between the grid and cathode is increased until the electrons have sufficient energy to cause ionization on collision with gas atoms (or molecules), e.g.

$$Ar(g) + e \text{ (fast)} \longrightarrow Ar^+(g) + 2e \text{ (slow)}$$

The positive ions are attracted to the anode and a current flows. The PD between the grid and cathode is recorded when this occurs and this is related to the *IE* by the following:

$$\text{1st } IE \text{ of gas} = 96{\cdot}3 \times \text{PD in V. kJ mol}^{-1}$$

IE values give information about the energy of the electron in the atom.

(a) Plot of successive *IE* values against the number of electrons removed

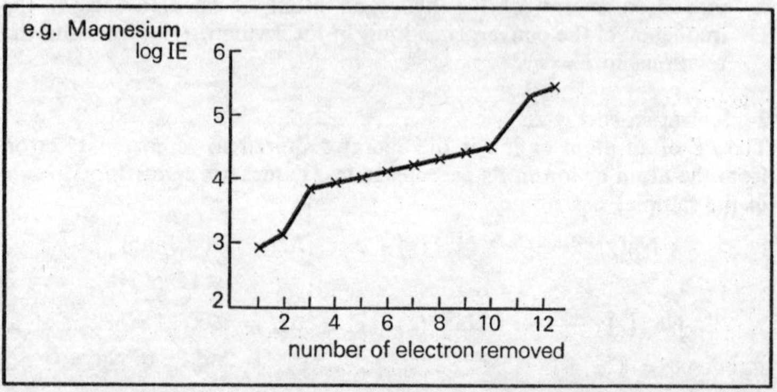

If all of the electrons had similar energies then the *IE* would be expected to increase regularly (the removal of each electron leaves a positive charge on the nucleus). However the graph indicates that the first two electrons are much more easily removed and so were in the highest energy level, i.e. 3*s*. The next 8 require more energy and occupy lower energy levels – the 2*s* and 2*p*. The 2 electrons in the 1*s* level are the most difficult to remove.

(b) Plot of the first *IE* of the elements against atomic number

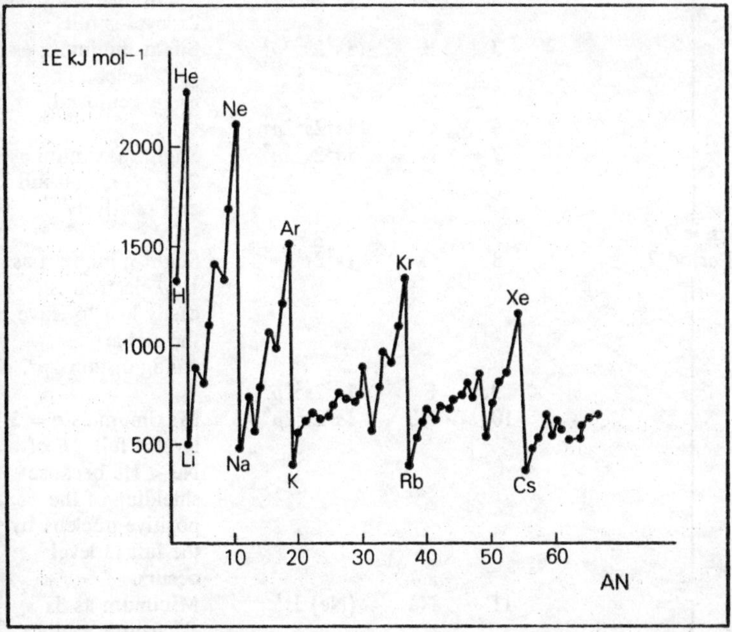

The graph indicates that:

(i) The first *IE* of the elements varies periodically with AN. This illustrates the Periodic Law – **the physical properties of the elements are a periodic function of their atomic number.**

(ii) The electrons with a particular *n* value do not have the same energy. When $n > 1$ the elements (ignoring the transition metals) fall into groups of 2, 3 and 3 with marked changes in *IE* values between the groups. This can be explained by a consideration of the electronic configuration of each element.

	AN	Element	Electronic configuration	Comments
$n = 1$ Period 1	1	H	$1s^1$	Electron relatively easily removed.
	2	He	$1s^2$	Maximum *IE* as 1*s* level is full.
$n = 2$ Period 2	3	Li	$1s^2 2s^1$	Minimum *IE* as $2s^1$ electron is easily removed.
	4	Be	$1s^2 2s^2$	Slight maximum as 2*s* level is full.
	5	B	$1s^2 2s^2 2p^1$	Slight minimum as $2p^1$ electron is easily removed.
	6	C	$1s^2 2s^2 2p^2$	
	7	N	$1s^2 2s^2 2p^3$	Slight maximum as 2*p* level is half full and relatively stable.
	8	O	$1s^2 2s^2 2p^4$	Slight minimum as 1 2*p* electron is easily lost to leave the stable configuration $2p^3$.
	9	F	$1s^2 2s^2 2p^5$	
	10	Ne	$1s^2 2s^2 2p^6$	Maximum as $n = 2$ level is full. *IE* of Ne $<$ He because shielding of the positive nucleus by the full 1*s* level occurs.
$n = 3$ Period 3	11	Na	(Ne) $3s^1$	Minimum as 3*s* electron is easily lost. *IE* Na $<$ Li because of shielding.
	12	Mg	(Ne) $3s^2$	Slight maximum. 3*s* level full.
	13	Al	(Ne) $3s^2 3p^1$	Slight minimum as 3*p* electron is easily lost.
	14	Si	(Ne) $3s^2\ 3p^2$	
	15	P	(Ne) $3s^2 3p^3$	Slight maximum as 3*p* level is $\frac{1}{2}$ full and relatively stable.

		AN	Element	Electronic configuration	Comments
		16	S	(Ne) $3s^2 3p^4$	Slight minimum as 1 of the $3p$ electrons is easily lost to leave it ½ full.
		17	Cl	(Ne) $3s^2 3p^5$	
		18	Ar	(Ne) $3s^2 3p^6$	Maximum as $3s$ and $3p$ are full.
		19	K	(Ar) $4s^1$	Minimum as $4s$ electron is easily removed.
		20	Ca	(Ar) $4s^2$	Slight maximum as $4s$ level is full.
		21	Sc	(Ar) $4s^2 3d^1$	
		22	Ti	(Ar) $4s^2 3d^2$	
		23	V	(Ar) $4s^2 3d^3$	
		24	Cr	(Ar) $4s^1 3d^5$	Stabler than $4s^2 3d^4$ as the $3d$ level is ½ full.
	1st Transitional Series	25	Mn	(Ar) $4s^2 3d^5$	
		26	Fe	(Ar) $4s^2 3d^6$	
		27	Co	(Ar) $4s^2 3d^7$	
		28	Ni	(Ar) $4s^2 3d^8$	
		29	Cu	(Ar) $4s^1$-$3d^{10}$	Stabler than $4s^2 3d^9$ as the $3d$ level is full.
$n = 4$ Period 4		30	Zn	(Ar) $4s^2 3d^{10}$	
		31	Ga	(Ar) $3d^{10}$-$4s^2 4p^1$	Slight minimum as the $4p$ electron is easily lost.
		32	Ge	(Ar) $3d^{10}$-$4s^2 4p^2$	
		33	As	(Ar) $3d^{10}$-$4s^2 4p^3$	Slight maximum as $4p$ level is ½ full and relatively stable.
		34	Se	(Ar) $3d^{10}$-$4s^2 4p^4$	Slight minimum as 1 of $4p$ electrons is easily lost.
		35	Br	(Ar) $3d^{10}$-$4s^2 4p^5$	
		36	Kr	(Ar) $3d^{10}$-$4s^2 4p^6$	Maximum as $4s$ and $4p$ levels are full.

The wave nature of the electron and atomic orbitals

In 1924 de Broglie suggested that the moving electron can be regarded as both a wave motion and as a particle. His equation relating these two aspects of the electron is

$$\lambda = \frac{\hbar}{\text{momentum of the electron}}$$

λ = wavelength; $\hbar$ = Planck's constant.

Experimental evidence for the wave nature of the electron was obtained by Davisson and Germer in 1927 when they diffracted an electron beam with a nickel crystal. Diffraction is a property of a wave motion and not of moving particles.

Schrödinger in 1927 evolved an equation which treated the electron as a 3-dimensional wave. From its solution for the 1 electron system, e.g. the hydrogen atom, it is possible to predict the probable position of an electron in the atom. The idea of a probability of finding an electron in a particular place is embodied in **Heisenberg's Uncertainty Principle.** This states that there is an uncertainty in defining the position and velocity of an electron at any given time as the electron is so small. Thus, any attempt to measure its velocity will alter its position and vice versa.

The probable position of the electron in its lowest energy state is represented by diagram (a). The density of dots represents the probability of finding the electron at a particular position.

a) variation of charge density for Is orbital

b) boundary surface for Is orbital

Diagram (b) shows the sphere in which the electron occupies 95% of its time, i.e. it is the boundary surface of the **1*s* orbital** of the atom.

All *s* orbitals are spherically symmetrical but the 2*s* differs from the 1*s* in that there is a greater probability of finding the electron further from the nucleus.

The *p* orbitals are dumb-bell in shape with a directional nature.

the p orbitals are dumb-bell in shape with a directional nature

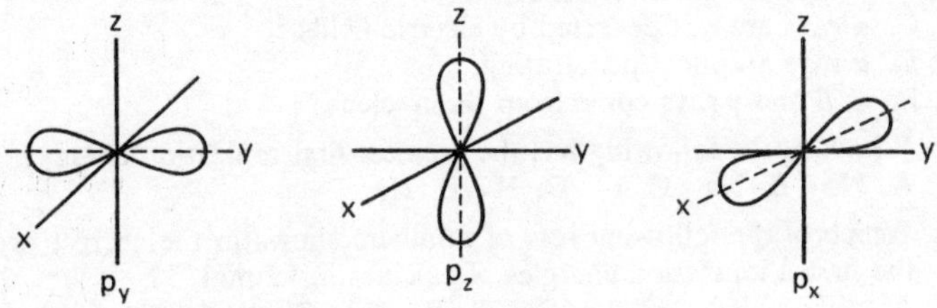

the boundary surfaces of the p_x, p_y and p_z orbitals

The *d* orbitals also have directional characteristics.

the d orbitals also have directional characteristics

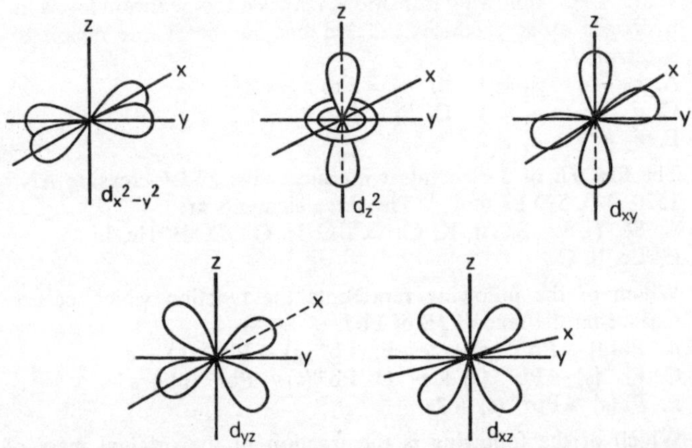

Practice questions

In questions 1 to 15 give the letter of the correct answer.

1 Which of the following species is obtained when $^{210}_{83}Bi$ emits an alpha and a beta particle?
A. $^{206}_{82}Pb$ B. $^{206}_{84}Po$ C. $^{208}_{80}Hg$ D. $^{206}_{81}Tl$ E. $^{207}_{82}Pb$

2 Which of the following must be emitted to produce $^{208}_{82}Pb$ from $^{224}_{88}Ra$?
A. 2α and 3β B. 3α and 3β C. 4α and 2β
D. 3α and 4β E. 4α and 3β

3 Which of the following statements about radiation is correct?
A. α rays have negligible mass.
B. γ rays are positively charged.
C. α rays are not deflected by electric fields.
D. α rays are most penetrating.
E. α, β and γ rays come from the nucleus.

4 Which of the following has the greatest first ionization energy?
A. Na B. Ne C. F D. H E. He

5 Which of the following sets of numbers shows in the correct order the first 4 ionization energies of calcium in kJ mol^{-1}?
A. 590 1100 4900 6500 B. 420 3100 4400 5900
C. 1260 2300 3800 5200 D. 580 1800 2700 11600
E. 790 1600 3200 4400

6 ^{32}P has a half-life of 14 days. The time taken for a given amount of ^{32}P to decay to $\frac{1}{4}$ of its original activity is
A. infinite time B. 14 days C. $3\frac{1}{2}$ days
D. 28 days E. 56 days

7 Which of the following transitions between the quantum levels in the hydrogen atom produces the 3rd line in the visible region of its spectrum?
A. $n_2 = 4$ to $n_1 = 1$ B. $n_2 = 5$ to $n_1 = 2$
C. $n_2 = 5$ to $n_1 = 1$ D. $n_2 = 3$ to $n_1 = 1$
E. $n_2 = 3$ to $n_1 = 2$

8 The first *IE* of 3 elements with successive and increasing AN are 1520, 420, 590 kJ mol^{-1}. The three elements are
A. Sc, Ti, V B. Ar, K, Ca C. C, N, O D. H, He, Li
E. Be, B, C

9 Which of the following represents the reaction which occurs in measuring the second *IE* of Pb?
A. $Pb(g) \rightarrow Pb^{2+}(g) + 2e$ B. $Pb^{+}(g) \rightarrow Pb^{2+}(g) + e$
C. $Pb^{+}(s) \rightarrow Pb^{2+}(s) + e$ D. $Pb^{+}(s) \rightarrow Pb^{2+}(g) + e$
E. $Pb(s) \rightarrow Pb^{2+}(s) + 2e$

10 Which of the following is the fraction of the original mass of a radioactive isotope of half-life 10 days which remains undecayed after 30 days?
A. $\frac{1}{30}$ B. $\frac{1}{3}$ C. $\frac{1}{4}$ D. $\frac{1}{2}$ E. $\frac{1}{8}$

Questions 11 to 15

Choose from A to E the electronic configuration of the elements in questions 11 to 15.
A. $1s^2 2s^2 2p^6 3s^1$ B. $1s^2 2s^2 2p^6 3s^2 3p^6 3d^2 4s^2$
C. $1s^2 2s^2 2p^1$ D. $1s^2 2s^2 2p^3$
E. $1s^2 2s^2 2p^6 3s^2 3p^6 3d^{10} 4s^2 4p^6$

11 A transition element

12 An alkali metal

13 A metal with maximum oxidation state +3

14 A noble gas

15 A non-metal with maximum oxidation state −3

16 (a) Explain the nature of α, β and γ emissions and compare their penetrating power and behaviour in magnetic fields.

(b) Identify *W*, *X*, *Y* and *Z* in the following equations:

i. $W \rightarrow {}^{219}_{86}\text{Rn} + \alpha$ ii. ${}^{14}_{7}\text{N} + {}^{1}_{0}n \rightarrow X + {}^{1}_{1}\text{H}$

iii. $Y + {}^{1}_{0}n \rightarrow {}^{3}_{1}\text{H} + \alpha$ iv. ${}^{228}_{88}\text{Ra} \rightarrow Z + \beta$

17 (a) Describe the essential features of the atomic spectrum of hydrogen and show how it leads to the concept of the principal energy levels in an atom.

(b) State how and why the first *IE* changes within Group 0 (the noble gases) and across Period 2.

18 The table gives the AN and MN of some stable elements and their isotopes.

AN	1	3,	6	8	10	16	19	30
MN	1, 2	7,	12, 13	16, 17	20, 21	33, 34	39, 40, 41	64, 67

Plot the numbers of neutrons against the corresponding numbers of protons and draw curves to define the lower and upper limits of stability. From the graph which of the following are radioactive?

${}^{42}_{19}\text{K}$ ${}^{31}_{15}\text{P}$ ${}^{59}_{27}\text{Co}$ ${}^{67}_{29}\text{Cu}$

2 Gases, Liquids and Solids

Gases and the kinetic theory

At high temperatures and low pressures, the behaviour of most gases is similar. Gases exhibiting this behaviour are said to be ideal and the laws applying to these gases are as follows.

1 *Boyle's law 1662*
The volume of a given mass of gas is universely proportional to the pressure at constant temperature.

i.e. $$p \propto \frac{1}{V}$$

or $$pV = \text{constant}$$

Thus, if V_1 is its volume at pressure p_1
and V_2 is its volume at pressure p_2

then $$p_1 V_1 = p_2 V_2 .$$

This law can be expressed graphically.

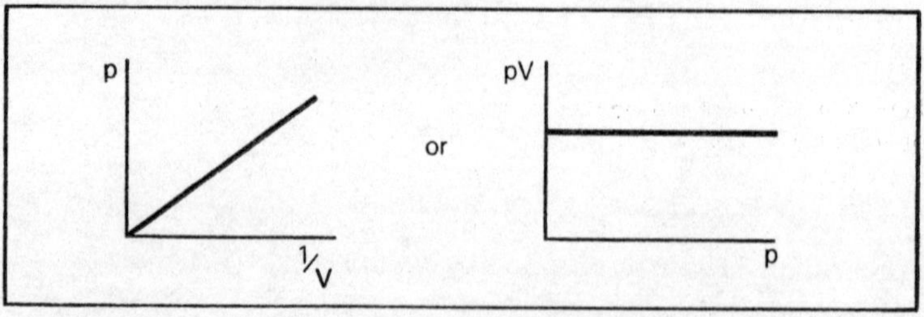

2 *Charles' law*
The volume of a given mass of gas is directly proportional to the absolute temperature at constant pressure.

i.e. $$V \propto T \qquad (T\,\text{K} = 273 + t°\text{C})$$

or $$V = \text{constant} \times T$$

Thus, if V_1 is its volume at temperature T_1
and V_2 is its volume at temperature T_2

then $$\frac{V_1}{T_1} = \frac{V_2}{T_2} .$$

This law can be expressed graphically.

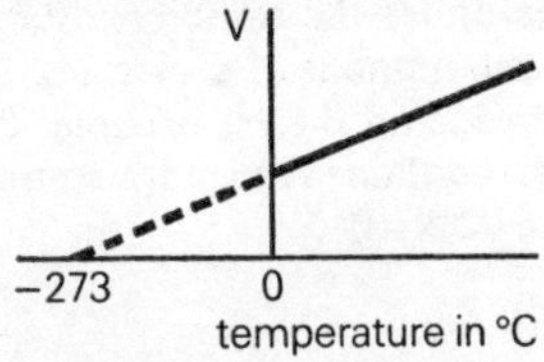

Theoretically a gas should have zero volume at 0 K (−273°C). However, in reality, gases either form liquids or solids before this temperature is reached.

3 *The ideal gas equation*

Boyle's and Charles' laws can be combined to give the ideal gas equation, i.e.

$$\frac{pV}{T} = \text{constant}$$

or

$$\frac{p_1V_1}{T_1} = \frac{p_2V_2}{T_2}.$$

If one mole of a gas is considered then

$$\frac{pV}{T} = R$$

where R is the universal gas constant or molar gas constant.
For n moles of gas

$$pV = nRT$$

$$n = \frac{\text{mass of gas}}{M_r \text{ in g}}$$

where M_r is the relative molecular mass of the gas.

Units (a) If p is in N m^{-2} or Pa (1 atm. = 101323 N m^{-2})
V is in m^3
T is in K
then $R = 8{\cdot}314$ J K^{-1} mol^{-1}.

(b) If p is in atm. (1 atm. = 760 mm of mercury)
V is in l
T is in K
then $R = 0{\cdot}0821$ l atm. K^{-1} mol^{-1}.

Standard temperature and pressure (s.t.p.) It is convenient to define a standard temperature and pressure for use when quoting volumes and densities of gases. A pressure of 1 atm. and a temperature of 0°C (273 K) have been accepted.

4 *Avogadro's law*

Equal volumes of gases under the same conditions of temperature and pressure contain the same number of molecules.
If 1 mole of a gas is considered then it occupies 22·4 l at s.t.p.
1 mole of any substance contains Avogadro's number of particles.
Avogadro's number = $6{\cdot}02 \times 10^{23}$

Gas densities

The density of a gas is expressed in 2 ways.

(a) Normal density ρ of a gas at a specified temperature and pressure is defined as

$$\rho = \frac{\text{mass}}{\text{volume}}$$

The density and M_r of the gas are related as follows

For n moles of the gas $pV = nRT$.

$$\text{But } n = \frac{\text{mass of gas}}{M_r \text{ in g}}$$

$$\therefore pV = \frac{\text{mass}}{M_r} RT.$$

$$\text{But } \rho = \frac{\text{mass}}{V}$$

$$\therefore p = \frac{\rho RT}{M_r}$$

[for a monatomic gas $p = \rho(RT/A_r)$].

For example at 273 K and $1{\cdot}01 \times 10^5$ N m^{-2}, $4{\cdot}46 \times 10^{-4}$ kg of argon occupy $2{\cdot}50 \times 10^{-4}$ m^3.
Find A_r of the gas.

$$\text{Density} = \frac{4{\cdot}46 \times 10^{-4}}{2{\cdot}50 \times 10^{-4}} \times 10^3 \text{ gm}^{-3}$$

But $$A_r = \rho \frac{RT}{p}$$

$$= \frac{4{\cdot}46 \times 10^{-4} \times 10^3 \times 8{\cdot}31 \times 273}{2{\cdot}50 \times 10^{-4} \times 1{\cdot}01 \times 10^5}$$

$$= 40$$

(b) Vapour density, VD, of a gas at a specified temperature and pressure is defined as

$$VD = \frac{\text{mass of a given volume of the gas}}{\text{mass of an equal volume of hydrogen under the same conditions}}$$

By Avogadro's law

$$VD = \frac{\text{mass of } n \text{ molecules of the gas}}{\text{mass of } n \text{ molecules of hydrogen}}$$

or $$VD = \frac{\text{mass of 1 molecule of the gas}}{\text{mass of 1 molecule of hydrogen}}$$

(M_r $H_2 = 2$)

$$\therefore \quad VD = \frac{M_r}{2}$$

or $$M_r = 2 \times VD$$

5 *Graham's law of gaseous diffusion*

The rate of diffusion (or effusion) of a gas at constant temperature and pressure is inversely proportional to the square root of its density. Diffusion provides evidence for the translational molecular motion of the gas particles.

$$\text{Rate of diffusion} \propto \frac{1}{\sqrt{\text{Density}}}$$

$$\text{but density } \rho = \frac{M_r p}{RT}$$

$$\therefore \text{ rate of diffusion} \propto \frac{1}{\sqrt{M_r}}$$

Thus for 2 gases A and B

$$\frac{\text{rate of diffusion of gas } A}{\text{rate of diffusion of gas } B} = \sqrt{\frac{M_r \text{ of } B}{M_r \text{ of } A}}$$

6 *Dalton's law of partial pressure*

The total pressure exerted by a mixture of gases is the sum of their partial pressures.

The partial pressure of a gas is the pressure it would exert if it alone occupied the total volume of the mixture at that temperature. For a mixture of gases A, B, C etc. of partial pressures p_A, p_B, p_C etc. the total pressure p of the mixture is given by

$$p = p_A + p_B + p_C + \cdots \text{ etc.}$$

where p_A = mole fraction of $A \times p$.

$$\text{Mole fraction of } A = \frac{\text{number of moles of } A}{\text{total number of moles}}$$

For example a globe of volume 2 m^3 containing oxygen at a pressure of 1×10^5 N m^{-2} was connected to a globe of volume 4 m^3 containing nitrogen at a pressure of $0{\cdot}25 \times 10^5$ N m^{-2} and the gases were allowed to mix. Calculate (a) the partial pressure of oxygen and (b) the total pressure assuming the temperature is constant.

(a) For the oxygen, using

$$p_1 V_1 = p_2 V_2$$

then
$$p_{O_2} = \frac{p_2 V_2}{V_1}$$
$$= \frac{1 \times 10^5 \times 2}{6}$$
$$= 3{\cdot}33 \times 10^4.$$

$\therefore$ Partial pressure of oxygen $= 3{\cdot}33 \times 10^4$ N m^{-2}

(b) For the nitrogen, using

$$p_1 V_1 = p_2 V_2$$

then
$$p_{N_2} = \frac{p_2 V_2}{V_1}$$
$$= \frac{0{\cdot}25 \times 10^5 \times 4}{6}$$
$$= 1{\cdot}67 \times 10^4$$

$\therefore$ Partial pressure of nitrogen $= 1{\cdot}67 \times 10^4$ N m^{-2}

$\therefore$ Total pressure $= (3{\cdot}33 \times 10^4) + (1{\cdot}67 \times 10^4)$
$= 5 \times 10^4$ N m^{-2}

Dalton's law is used to find the partial pressure of a gas when it is collected over water.
Atmospheric pressure = partial pressure of gas + saturated water vapour pressure at the temperature of collection.
Tables of the saturated vapour pressure of water at various temperatures are available.

The kinetic theory

The Kinetic Theory leads to a mathematical relationship between the pressure of a gas and the average kinetic energy of its molecules.

Postulates

1 Matter is made up of small particles – molecules or atoms.

2 The particles are in continual rapid, random translational motion in straight lines. The bombardment of the particles on the vessel walls gives rise to the pressure of the gas.

3 The particles are very far apart compared with their size. Thus the forces of attraction between the particles and their actual volume are negligible.

4 The collisions are perfectly elastic. Therefore the average and the total kinetic energy of the gas is constant at constant temperature.

From these postulates it can be shown that for n particles of gas each of mass m,

$$pV = \tfrac{1}{3}mnc^2$$

where p is the pressure; V is the volume; c^2 is the mean square velocity.

$$c = \sqrt{\frac{c_1{}^2 + c_2{}^2 + c_3{}^2 \cdots c_n{}^2}{n}}$$

$c_1, c_2 \ldots c_n$ are the velocities of the individual particles.

For 1 mole of gas

$$\begin{aligned} pV &= \tfrac{1}{3}M_r c^2 \\ &= RT \end{aligned}$$

where M_r is the relative molecular mass of the gas

and
$$\begin{aligned} \text{total } KE &= \tfrac{1}{2}M_r c^2 \\ &= \tfrac{3}{2}RT \end{aligned}$$

Thus the total, and therefore the average, kinetic energy of the molecules is directly proportional to the absolute temperature.

Proof

Consider a cube of side l containing n particles each of mass m with velocities $c_1, c_2, c_3, \ldots, c_n$.

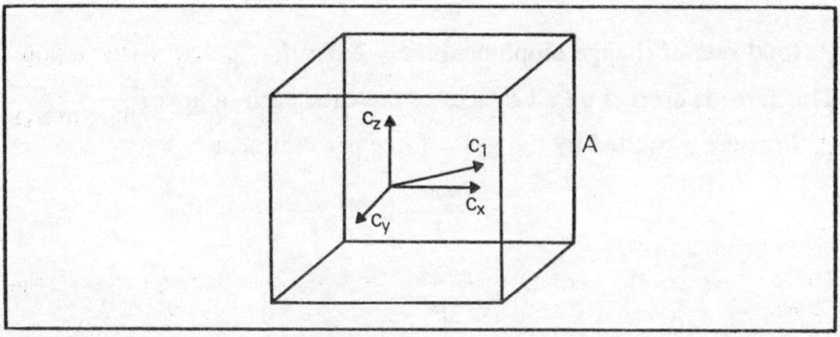

Consider 1 particle with velocity c_1. Its velocity can be resolved into 3 components parallel to the edges of the cube. Thus,

$$c_1^2 = c_x^2 + c_y^2 + c_z^2$$

The momentum of the particle relative to the x axis $= mc_x$. The change in momentum of the particle relative to the x axis on striking the wall A

$$= mc_x - (-mc_x)$$
$$= 2mc_x.$$

The time taken for it to rebound from the opposite wall and hit A again

$$= \frac{2l}{c_x}.$$

∴ Rate of change of momentum relative to the x axis

$$= (2mc_x + 2mc_x)\frac{c_x}{2l}$$
$$= \frac{2mc_x^2}{l}.$$

Similarly the rate of change of momentum relative to the y and z axes is $2mc_y^2/l$ and $2mc_z^2/l$ respectively.

∴ Total rate of change of momentum

$$= \frac{2mc_x^2}{l} + \frac{2mc_y^2}{l} + \frac{2mc_z^2}{l}$$
$$= \frac{2mc_1^2}{l}.$$

For all the particles the total rate of change of momentum

$$= \frac{2m}{l}(c_1^2 + c_2^2 + c_3^2 \cdots c_n^2)$$

but $$c^2 = \frac{c_1^2 + c_3^2 \cdots c_n^2}{n}$$

∴ total rate of change of momentum $= 2mnc^2/l$.

This force is exerted on all 6 faces of the cube each of area l^2.

∴ Pressure p exerted by the gas = force per unit area

$$= \frac{2mnc^2}{l} \div 6l^2$$
$$= \frac{mnc^2}{3l^3}$$

but $V = l^3$ where V is the volume of the cube

$$\therefore p = \frac{1mnc^2}{3V}$$

$$\therefore pV = \tfrac{1}{3}mnc^2.$$

For 1 mole of the gas of relative molecular mass M_r,

$$pV = \tfrac{1}{3}M_r c^2 = RT.$$

$$\text{Total } KE = \tfrac{1}{2}M_r c^2 = \tfrac{3}{2}RT.$$

$\therefore$ The total KE and the average $KE \propto$ absolute temperature. These equations can be used to deduce the laws relating to ideal gases.

(a) **Boyle's law**

$$\text{Total } KE = \tfrac{1}{2}mnc^2$$

$$\text{since } pV = \tfrac{1}{3}mnc^2$$

$$pV = \tfrac{2}{3} \text{ total } KE.$$

But the total KE is constant at constant temperature.

$$\therefore p \propto \frac{1}{V}.$$

(b) **Charles' law**

$$\text{Total } KE \propto T$$

$$\text{i.e. } \tfrac{1}{2}mnc^2 \propto T$$

$$\therefore \tfrac{1}{3}mnc^2 \propto T$$

$$\text{i.e. } pV \propto T.$$

If p is constant then $V \propto T$.

(c) **Avogadro's hypothesis** Consider 2 gases of equal volume at the same temperature and pressure. Then for the 1st gas

$$pV = \tfrac{1}{3}m_1 n_1 c_1{}^2$$

and for the 2nd gas

$$pV = \tfrac{1}{3}m_2 n_2 c_2{}^2.$$

Since the gases are at the same temperature their average KE are the same. Then,

$$\tfrac{1}{2}m_1 c_1{}^2 = \tfrac{1}{2}m_2 c_2{}^2.$$

Then n_1 must equal n_2. This is a statement of Avogadro's hypothesis.

(d) **Graham's law**

$$pV = \tfrac{1}{3}mnc^2$$

$$\therefore c^2 = 3\frac{pV}{mn}.$$

But density $$\rho = \frac{mn}{V}$$

$$\therefore c^2 = \frac{3p}{\rho}$$

or $$c = \sqrt{\frac{3p}{\rho}}.$$

At constant pressure

$$c \propto \frac{1}{\sqrt{\rho}}$$

However the rate of diffusion of a gas is proportional to c.

$\therefore$ Rate of diffusion = $1/\sqrt{\rho}$.

(e) **Dalton's law** For a gas $pV = \frac{2}{3}$ total KE.
If a mixture of 2 gases A and B is considered then

$$\text{total } KE = E_A + E_B$$

where E_A and E_B are the KE of gases A and B.

$$\therefore pV = \tfrac{2}{3}(E_A + E_B).$$

If the pressure exerted by A is p_A, i.e. its partial pressure, then

$$p_A V = \tfrac{2}{3}E_A,$$

and for gas B,

$$p_B V = \tfrac{2}{3}E_B.$$

It therefore follows that

$$p = p_A + p_B.$$

Molecular velocities
The velocities of the individual particles in a gas vary. The root mean square velocity c of the gas may be calculated from the density.

For example the density of hydrogen at s.t.p. = 0·09 kg m^{-3} but

$$c = \sqrt{\frac{3p}{\rho}} = \sqrt{\frac{3 \times 101325}{0{\cdot}09}}$$
$$= 1838 \text{ m s}^{-1}.$$

Hydrogen has the lowest density and therefore the highest root mean square velocity of any gas at a particular temperature and pressure.

Distribution of molecular velocities

The molecules in a gas undergo collisions and so their velocities vary. The distribution of molecular velocities is investigated using Zartmann's apparatus.

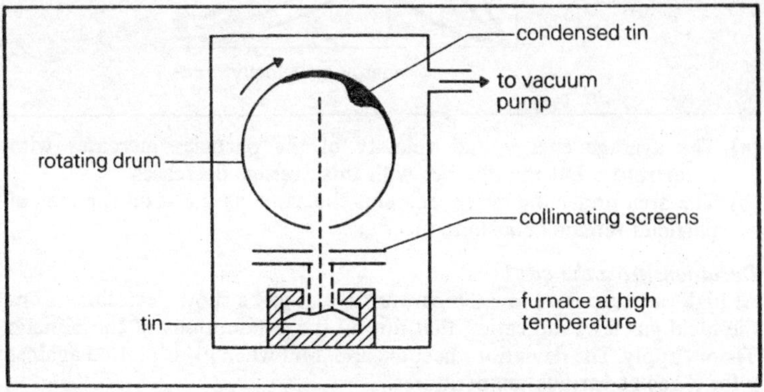

The sample of tin is atomized in the furnace and the stream of tin atoms emerge through the collimating screens into the evacuated system. They enter the rotating drum when the slit is opposite the holes in the collimating screens. The atoms travelling most quickly reach the back of the drum and condense first. The slower atoms arrive later and condense at a different place. After several rotations the thickness of the condensed tin is measured at different points. Knowing the dimensions of the apparatus, and the speed of rotation, the fraction of molecules with each velocity is calculated. The experiment is repeated at different furnace temperatures.

By applying statistical techniques to the Kinetic Theory, Boltzmann in 1886 deduced an equation which enabled the number of molecules in a gas with an energy *E* or greater to be calculated. The equation is

$$n = n_0\, e^{-E/RT}$$

where n_0 is the total number of molecules; *R* is the gas constant; *T* is the absolute temperature.

The results are represented graphically for a gas at temperatures T_1, T_2 and T_3 where $T_3 > T_2 > T_1$. Similar curves describe the distribution of molecular velocities in a gas.

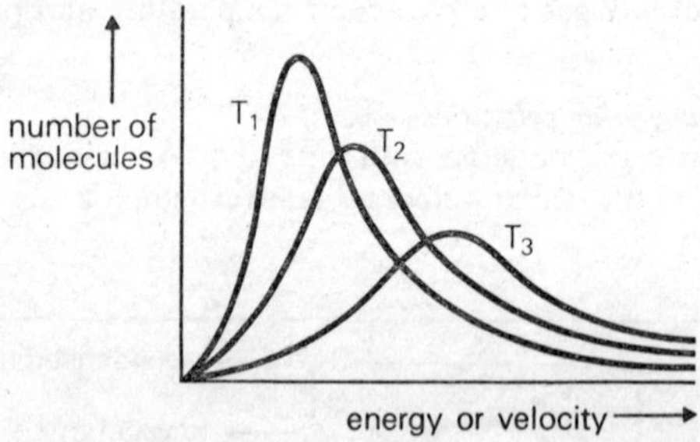

(a) The average energy and velocity of the particles increases with temperature but the number with this average decreases.
(b) The area under the curves remains the same as the total number of particles remains constant.

Deviations from the gas laws
At high pressures and low temperatures, real gases show deviations from the ideal gas laws indicating that not all the assumptions of the Kinetic Theory apply. The deviations become apparent when pV is plotted against p for a gas at various temperatures.

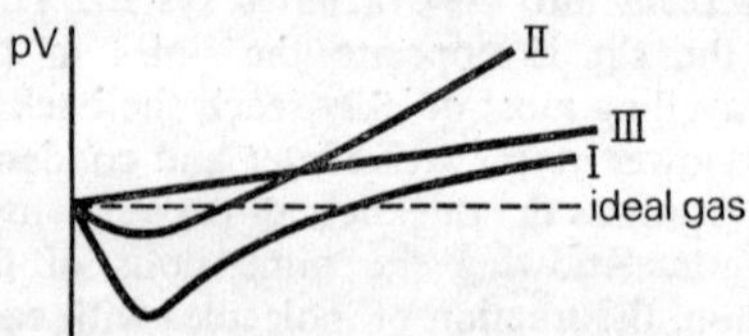

I at low temperatures (i.e. temperatures near its b.pt.) the gas is at first more compressible than expected and then less so

II at slightly higher temperatures there is still a slight minimum but it is less pronounced

III at temperatures of at least 300°C above the bpt. the gas is less compressible than expected

At 40°C, gases showing curves similar to I, II and III are carbon dioxide, oxygen and hydrogen respectively.

In the Kinetic Theory it is assumed that the volume of the particles themselves and the forces between them are negligible. However, at low temperatures and high pressures, when the particles are close together and moving slowly, these assumptions become invalid. In 1873 van der Waals introduced a modified version of the general gas equation to account for these deviations.

$\left(p + \frac{a}{V_2}\right)(V - b) = nRT$ (a and b are constants for a particular gas).

The term a/V^2 accounts for the attractive forces between the particles which causes the pressure exerted by the gas to be less than the ideal value. These forces are known as van der Waals' forces. The constant b accounts for the volume of the particles themselves and b becomes more significant as V becomes small, i.e. when the gas is highly compressed.

This equation describes the behaviour of real gases more satisfactorily than the ideal gas equation.

Determination of the relative molecular mass of gases and volatile liquids

In all of the following methods (except that of limiting densities) it is assumed that the gas involved does not associate or dissociate and that it behaves ideally under the conditions of the experiment.

Gases

1 **Diffusion measurements** The rate of diffusion of a gas can be compared with that of a gas of known M_r. Then by Graham's law,

$$\frac{R_1}{R_2} = \sqrt{\frac{M_{r2}}{M_{r1}}}$$

R_1 and R_2 are the rates of diffusion of gases 1 & 2 of relative molecular masses M_{r1} and M_{r2}.

2 **Density measurements** A syringe is weighed containing a known volume of air and then reweighed containing the same volume of gas. Knowing the density of air is 1·293 g l^{-1} at s.t.p., M_r of the gas is found using the relationship

$$\text{density} = \frac{\text{mass}}{\text{volume}} = \frac{M_r p}{RT}$$

3 **Method of limiting densities** This method enables M_r values of gases to be calculated more accurately as it takes into account deviations

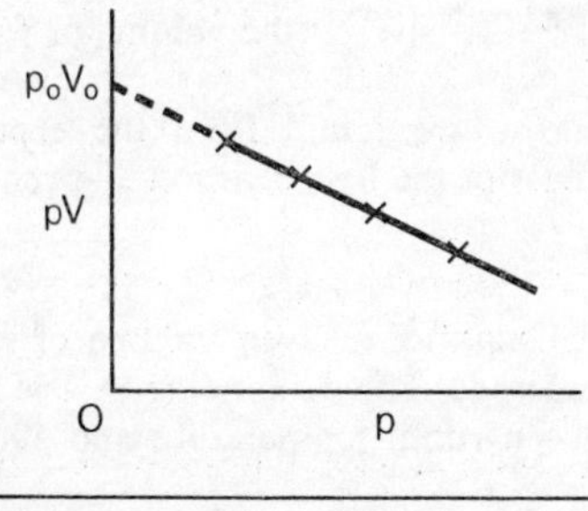

from ideal behaviour. The volume of a given mass of gas at 0°C is found at various pressures. A graph is plotted of pV against p and the line is extrapolated back to zero pressure where the gas obeys the gas laws. The value of $p_0 V_0$ obtained by extrapolation is then used for the term pV in the equation for finding M_r.

$$pV = \frac{\text{mass of gas}}{M_r} RT,$$

e.g. for 1 g of sulphur dioxide at 0°C,

$$p_0 V_0 = 0{\cdot}3488 \text{ atm. l.}$$

$$\therefore 0{\cdot}3488 = \frac{1}{M_r} \times 0{\cdot}0821 \times 273$$

$$M_r = \frac{0{\cdot}0821 \times 273}{0{\cdot}3488}$$

$$= 64{\cdot}3$$

M of sulphur dioxide = 64·3.

4 **Eudiometry** (For gaseous hydrocarbons) A known volume of the gaseous hydrocarbon is mixed with a measured excess of oxygen in a eudiometer and exploded by sparking. On cooling, the water vapour formed condenses and its volume is negligible. The decrease in volume after sparking and cooling to room temperature is noted. Potassium hydroxide solution is added to absorb the carbon dioxide and the final volume recorded.
If the formula of the hydrocarbon is $C_x H_y$, then

$$C_x H_y + \left(x + \frac{y}{4}\right) O_2 \longrightarrow xCO_2 + \frac{y}{2} H_2O.$$

By Avogadro's law

$$1 \text{ vol} + \left(x + \frac{y}{4}\right) \text{volumes} \longrightarrow x \text{ volumes}$$

(the volume of water is negligible)

The values of x and y are found from the experiment and the relative molecular mass of the hydrocarbon is given by $(12x + y)$.

Volatile liquids

1 **Dumas' method** The mass of a given volume of vapour from the volatile liquid is found under known conditions. The liquid must have a boiling point between room temperature and 100°C, e.g. ethoxyethane.

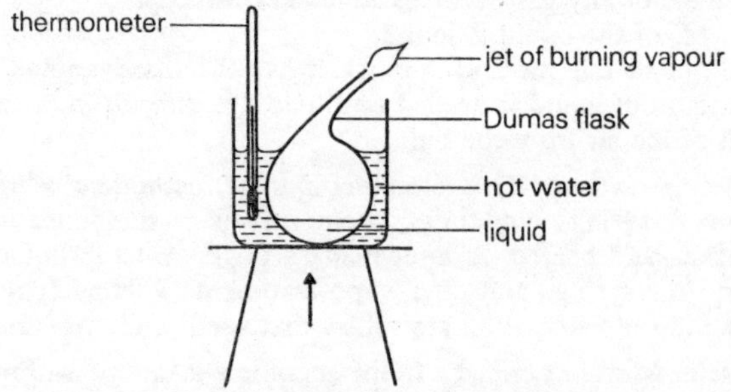

(a) The bulb is weighed empty, i.e. full of air (a g).
(b) A small volume of the liquid is introduced into the flask by warming it and then letting it cool with the glass jet under the liquid surface.
(c) It is placed in the water bath as shown. The temperature of the water T_b is kept constant at about 30°C above the b.pt. of the liquid. The vapour is burnt at the jet until all of the liquid has disappeared. The bulb full of vapour is then sealed and reweighed (b g).
(d) The jet is broken under the surface of cold water (previously boiled) which is sucked up to fill the bulb. The bulb full of water is reweighed (c g). Room temperature and pressure are recorded.

Calculation

Mass of water in the bulb = $(c - a)$ g
∴ internal volume of the bulb = $(c - a)$ cm³.
The volume of air inside the bulb is converted to s.t.p. using $p_1V_1/T_1 = p_2V_2/T_2$.
Let the volume of air at s.t.p. be x cm³.

$$\begin{aligned}
&\text{Density of air at s.t.p.} &&= 0{\cdot}0013 \text{ g cm}^3.\\
&\therefore \text{Mass of air in the bulb} &&= (0{\cdot}0013 \times x) \text{ g}.\\
&\therefore \text{Mass of evacuated bulb} &&= (a - 0{\cdot}0013x) \text{ g}.\\
&\therefore \text{Mass of vapour in the bulb} &&= [b - (a - 0{\cdot}0013x)] \text{ g}\\
& &&= y \text{ g}.
\end{aligned}$$

But this mass of vapour (y g) occupies a volume of $(c - a)$ cm³ at atmospheric pressure and the temperature of the water bath T_b°C. This volume is converted to s.t.p. (z cm³).
Thus y g of the liquid would theoretically occupy z cm³ as a gas at s.t.p.

By Avogadro's law
1 mole of any gas occupies 22400 cm^3 at s.t.p.
∴ M_r of the liquid is found.
Although this method is quick it has the disadvantage that a large amount of liquid is needed and that it is difficult in practice to expel all of the air from the bulb.

2 **Syringe method** This resembles Dumas' method but is easier to carry out. A syringe containing 5 cm^3 of air is surrounded by a steam jacket and heated. A known mass of the volatile liquid is injected into the syringe and after vaporization its volume is recorded. The method of calculation resembles that used in Dumas' method.

3 **Victor Meyer's method** In this method a given mass of the liquid *x* is vaporized and the volume it would occupy as a gas at room temperature and pressure is found by displacing the equivalent volume of air.

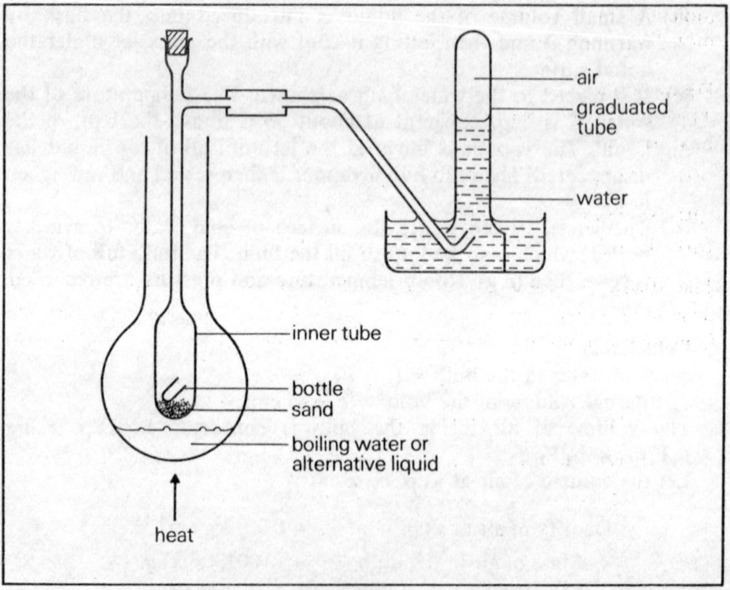

(a) The liquid in the outer jacket with a b.pt. of at least 30°C above that of the liquid *x* is heated until equilibrium is reached, i.e. no more air is displaced from the inner container.
(b) A small bottle is weighed empty and almost full of liquid *x*. The mass (*a* g) of liquid *x* is calculated.
(c) The bottle is dropped into the inner tube and the bung quickly replaced. The liquid vaporizes and the vapour displaces its own volume of air which is collected by displacement of water. The

volume of air obtained V_x cm³ is measured after the water levels inside and outside the tube have been equalized. Atmospheric pressure (p_x mm of Hg) and room temperature are recorded.

Calculation

Saturated water vapour pressure at room temperature = y mm of Hg.
$\therefore$ Pressure of dry air displaced = $(p_x - y)$ mm Hg.
The volume of air displaced at s.t.p. (z cm³) is found using $p_1 V_1/T_1 = p_2 V_2/T_2$.
a g of liquid would theoretically occupy z cm³ at s.t.p. as a gas. Since 1 mole of any gas occupies 22400 cm³ at s.t.p. the relative molecular mass of the liquid is found. For this method to be accurate the liquid must vaporize rapidly.

Dissociation and association

If dissociation or association produces a change in the number of gaseous molecules present then the measured vapour density and therefore relative molecular mass will be different from that calculated from the formula.

Dissociation

Thermal dissociation is the reversible decomposition of a substance on heating, e.g.

(i) $2HI(g) \rightleftharpoons H_2(g) + I_2(g)$
(ii) $N_2O_4(g) \rightleftharpoons 2NO_2(g)$
(iii) $PCl_5(g) \rightleftharpoons PCl_3(g) + Cl_2(g)$

The degree of dissociation, α, is the fraction or percentage of the originally undissociated molecules which have undergone dissociation. Consider 1 mole of a compound X which dissociates into n molecules

$$X \rightleftharpoons A + B + \cdots n$$

No. of moles before dissociation 1 0

No. of moles after dissociation $(1 - \alpha)$ $n\alpha$

$$\frac{\text{Total no. of moles after dissociation}}{\text{No. of moles before dissociation}} = \frac{1 - \alpha + n\alpha}{1} = \frac{1 + \alpha(n-1)}{1}.$$

By Avogadro's law, if the number of particles increases then the volume of the gas increases. The VD therefore decreases.

$$\therefore \quad \frac{VD \text{ before dissociation}}{VD \text{ after dissociation}} = \frac{1 + \alpha(n-1)}{1}$$

or $$\frac{M_r \text{ calculated (assuming no dissociation)}}{M_r \text{ actual (from experiment)}} = \frac{1 + \alpha(n-1)}{1}.$$

For example find the degree of dissociation of PCl_5 at 200°C if the *VD* at this temperature is 70.

$$PCl_5 \rightleftharpoons PCl_3 + Cl_2$$

	PCl_5	PCl_3	Cl_2
No. of moles before dissociation	1	0	0
No. of moles after dissociation	$(1-\alpha)$	α	α

$$\frac{\text{Total no. of moles after dissociation}}{\text{No. of moles before dissociation}} = \frac{1+\alpha}{1}$$

$$= \frac{VD \text{ before dissociation}}{VD \text{ after dissociation}}$$

$$M_r\ PCl_5 = 208{\cdot}5$$

$$\therefore \quad VD = 104{\cdot}25$$

$$\therefore \quad \frac{1+\alpha}{1} = \frac{104{\cdot}25}{70}$$

$$\therefore \quad \alpha = 0{\cdot}489 \text{ or } 48{\cdot}9\,\%$$

The degree of dissociation of PCl_5 at 700°C is 48·9 %.

Association

Association is the reversible combination of 2 or more like molecules to form a single molecule.

Association of gaseous molecules produces a decrease in volume and thus a corresponding increase in the *VD* and M_r values.
For example at certain temperatures in the vapour state $AlCl_3$, $FeCl_3$ and CH_3COOH exist as dimers.

Critical constants

When a real gas is cooled below a certain temperature called its critical temperature, T_c, and then compressed, it liquefies. T_c is the temperature above which it is impossible to liquefy a gas by compression alone. The critical pressure, p_c, is the pressure needed to liquefy a gas at its T_c.

The critical volume V_c, is the volume occupied by 1 mole of the gas at its T_c and p_c.

A **permanent** gas has a critical temperature which lies below room temperature. Therefore, at room temperature, it cannot be liquefied by pressure alone. For example oxygen, nitrogen, hydrogen, carbon monoxide.

Andrews, in 1869, investigated the behaviour of carbon dioxide in the region of its critical constants. He obtained values of pressure and volume of a given mass of the gas at different temperatures. The values of pressure were plotted against values of volume at each temperature to give a series of isothermals.

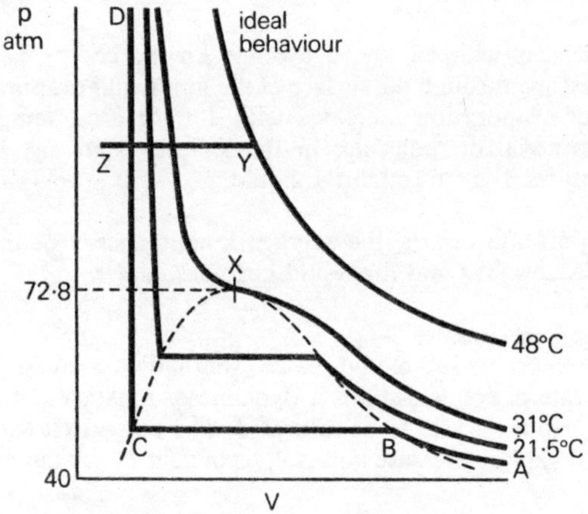

At 48°C and above, carbon dioxide obeys Boyle's law.

At temperatures below 31°C the isothermals are discontinuous.

Consider curve *ABCD*.

AB – effect of an increase in pressure on the volume of gas.

BC – liquefaction occurs. There is a large volume change with no pressure change.

CD – effect of pressure on the volume of a liquid. Liquids are difficult to compress.

The end of the horizontal portions of different isothermals form an approximate parabola with the critical point *X* at its apex.

Above the T_c, the isothermals show no discontinuity so it is not possible to detect the precise point at which a liquid becomes a gas and vice versa. If a liquid represented by point *Z* is heated at constant pressure to 48°C its condition at different temperatures is represented by line *ZY*. At *Y* it is a gas. The change occurs smoothly and continuously and is described by the term 'continuity of state'.

Liquids

The particles in a liquid undergo rapid random motion (Brownian Motion) but are closer together and move more slowly than in a gas.

The cohesive forces between the particles are therefore more significant. At the liquid surface the particles experience a net downward force giving rise to the surface tension.

Evaporation

Some molecules with an above average kinetic energy have sufficient energy to escape through the surface of the liquid and evaporation occurs. The rate of evaporation increases with surface area, temperature and with the removal of molecules in the gaseous state by, for example, bubbling an inert gas through the liquid.

When evaporation occurs, the average kinetic energy of the remaining liquid molecules falls and the liquid becomes cooler.

Saturated vapour pressure

If evaporation occurs in a closed vessel, eventually the rate of evaporation equals the rate of condensation. A dynamic or kinetic equilibrium is set up and the vapour is said to be saturated. The pressure it exerts is called its saturated vapour pressure and is independent of the amount of liquid present.

The saturated vapour pressure increases exponentially with temperature and when it equals atmospheric pressure the liquid boils, i.e. vaporization occurs throughout the whole volume of the liquid. The b.pt. remains constant until all the liquid has boiled away.

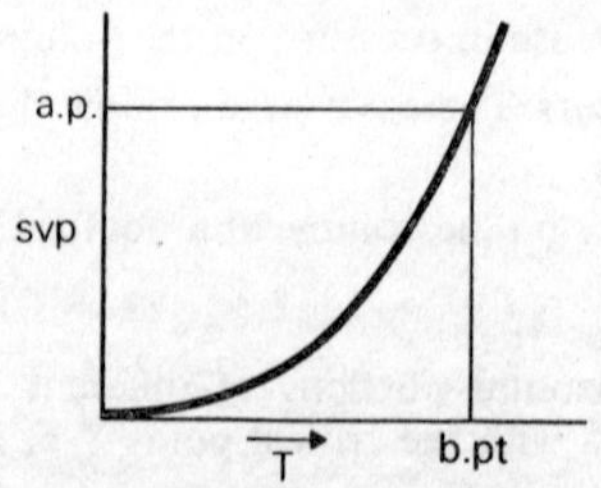

The equation connecting the effect of the temperature T_1 on the saturated vapour pressure, p, is given by

$$\log_e p = -\frac{\Delta H_{\text{vap}}}{RT} + \text{constant}.$$

Where ΔH_{vap} is the molar latent heat of evaporation, or the enthalpy of evaporation, and R is the universal gas constant. Or

$$\log_{10} p = -\frac{\Delta H_{\text{vap}}}{2{\cdot}303RT} + \text{constant}.$$

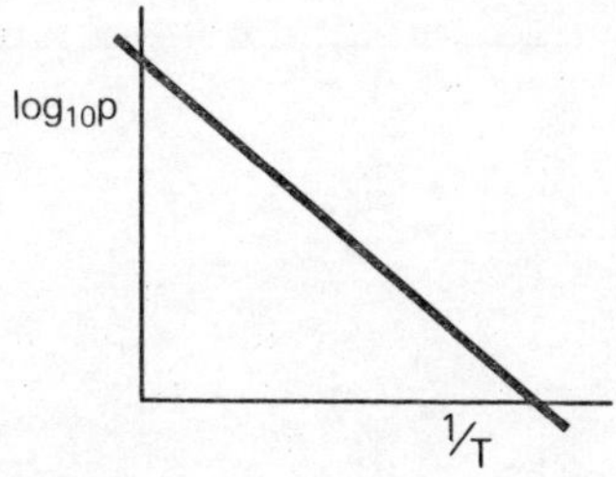

On plotting $\log_{10} p$ against $1/T$ a straight line of gradient $-\Delta H_{vap}/2{\cdot}303R$ is obtained. ΔH_{vap} can therefore be found.

The enthalpy of evaporation is the amount of heat required to change 1 mole of a liquid at its boiling point to a gas at the same temperature. Its value depends on the relative molecular mass of the substance as well as on the nature of the intermolecular forces.

Influence of shape of molecules and intermolecular forces on b.pts.

Examples

(a) C_4H_{10}

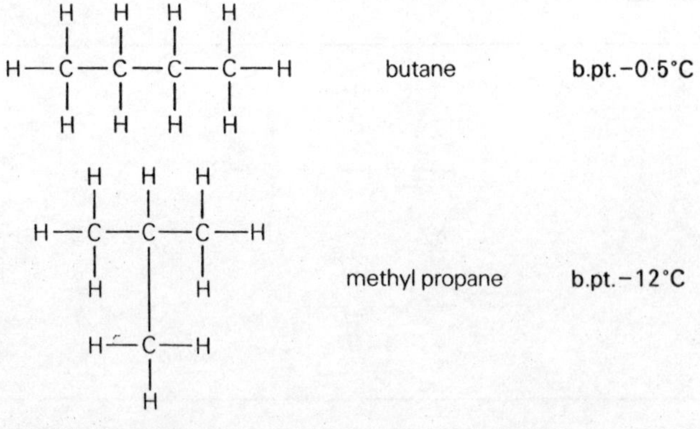

The more symmetrical isomer has the lower b.pt. as there is less intermolecular interference.

(b) $C_2H_4O_2$

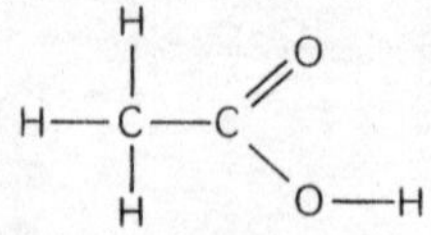

ethanoic acid b.pt. 118°C

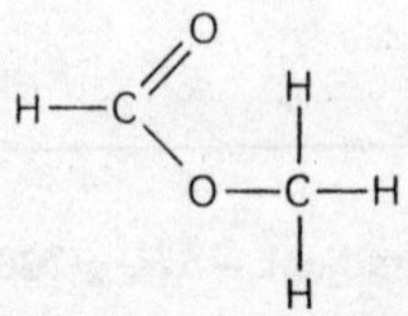

methyl methanoate b.pt. 32°C

Acids have higher b.pts. than esters as hydrogen bonding exists between the acid molecules.

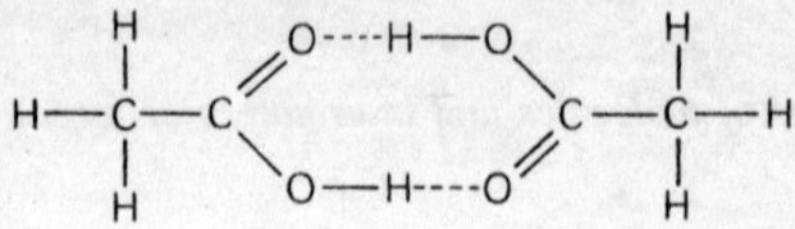

Electrostatic interaction is present between ester molecules, due to the polarization of the carbonyl group but this is weaker than hydrogen bonding.

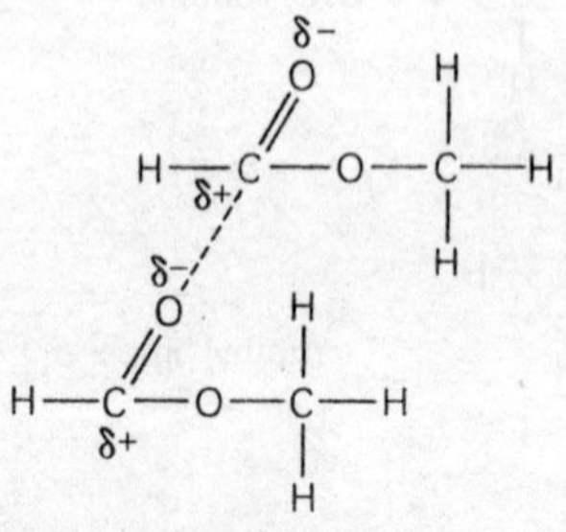

This electrostatic interaction also occurs between molecules of aldehydes and ketones so giving them higher b.pts. than alkanes or alkenes of the same relative molecular mass.

Solids

In solids the particles (atoms, molecules or ions) are kept in a regular, orderly array by cohesive forces. They are closer together than in a liquid and their motion is limited to vibration or oscillation about a fixed point. When a solid is heated the vibration increases until the cohesive forces can no longer keep the particles together and the solid melts.

The molar latent heat of fusion (enthalpy of fusion) is the heat required to change 1 mole of a solid to a liquid at its melting point. Its value gives an indication of the type of cohesive forces between the particles in the solid.

Practice questions

In questions 1 to 10 give the letter of the correct answer.

1 Which of the following is the volume in cm^3 occupied by 30 cm^3 of gas after its temperature has been raised from 27°C to 127°C and its pressure reduced from 760 mm of Hg to 380 mm of Hg?
A. 20 B. 45 C. 60 D. 80 E. 282

2 Which of the following is the ratio of the mean kinetic energies of the molecules in 3 identical vessels at 298 K containing 8 moles of hydrogen 1 mole of nitrogen and 3 moles of chlorine respectively?
A. 1 : 1 : 1 B. 8 : 14 : 71 C. 8 : 1 : 3 D. $\frac{1}{8}$: 1 : $\frac{1}{3}$ E. 2 : 28 : 71

3 Which of the following statements about the properties of the molecules in a mixture of oxygen and nitrogen must be correct?
A. The mean velocities are equal.
B. The velocities are independent of temperature.
C. The mean kinetic energies are equal.
D. The molecular volumes are equal.
E. The oxygen molecules move faster than the nitrogen molecules.

4 The enthalpies of vaporization of benzene and water are 30·7 and 40·6 kJ mol^{-1} respectively. This is because
A. benzene has a larger relative molecular mass than water.
B. benzene is a polar solvent.
C. hydrogen bonding exists between water molecules.
D. electrostatic interaction exists between benzene molecules.
E. the water molecules move faster than the benzene molecules.

5 When the number of molecules with a given velocity in a gaseous sample is plotted against velocity, a 'distribution' curve is obtained. If the temperature of the gas is decreased:
A. The peak of the distribution is moved to a lower velocity.
B. The width of the distribution is increased.
C. The maximum height of the distribution is decreased.

D. The area under the curve is decreased.
E. The number of molecules with infinite velocity becomes zero.

6 2 gases *A* and *B* have M_r values of 64 and 16 respectively. Which of the following is the ratio of the relative rates of diffusion of *A* and *B*?

A. 1 : 1 B. 1 : 2 C. 1 : 4 D. 2 : 1 E. 4 : 1

7 18 g of glucose (M_r 180) are dissolved in 2·4 moles of water. Which of the following is the mole fraction of the glucose in the resulting solution?

A. 0·04 B. 0·05 C. 0·1 D. 7·5 E. 18

8 Ethane burns in oxygen according to the equation

$$2C_2H_6(g) + 7O_2(g) \longrightarrow 4CO_2(g) + 6H_2O(g)$$

If 4 l of ethane burn in 20 l of oxygen at 1 atm and 4000 K, what is the final volume in l of the gaseous mixture under these conditions?

A. 4 B. 10 C. 16 D. 19 E. 26

9 When 0·224 l (at s.t.p.) of an organic compound were completely combusted in excess oxygen, 0·08 mole of carbon dioxide were formed. How many carbon atoms are there in 1 molecule of the compound?

A. 1 B. 2 C. 4 D. 8 E. 80

10 The boiling points of two isomeric compounds, propanoic acid and methyl ethanoate, are 141°C and 57°C respectively. Which of the following best accounts for this difference?

A. Propanoic acid has a higher M_r.
B. Propanoic acid ionizes, the ester does not.
C. Hydrogen bonding exists between the acid molecules.
D. Van der Waals' forces exist between the ester molecules.
E. The ester decomposes at a lower temperature.

11 Outline a method of determining the relative molecular mass of a volatile liquid. In what circumstances will this method produce abnormal results? A compound of phosphorus and fluorine contains 75·4% of fluorine. 2·00 g of the compound has a volume of 389 cm^3 at 1 atm and 25°C. Calculate M_r of the compound and find its formula (A_r F = 19, P = 30·97, molar volume of a gas is 22·4 l at s.t.p.).

12 The following table shows the variation of volume with pressure for 1 g of oxygen at 0°C.

p/atm	2·00	4·00	6·00	9·00
V/litre	0·34955	0·17442	0·11607	0·07714

(i) Plot a graph of pV against p.

(ii) On the same diagram show with a dotted line the relationship between pV and p if in the same experiment, the gas had behaved ideally.

(iii) Use the graph to calculate the value for the gas constant. (A_r O = 16)

13 The equation relating the vapour pressure of a liquid to the absolute temperature T and to the enthalpy of vaporization ΔH_{vap} is

$$2{\cdot}303 \log_{10} p = -\frac{\Delta H_{\text{vap}}}{RT} + \text{constant.}$$

The table lists the vapour pressure of water in mm of Hg at various temperatures in K.

Temp/K	Reciprocal Temp K^{-1}	p/mm Hg	$\log_{10} p$/mm Hg
342	0·002924	223·9	2·350
353	0·002833	355·1	2·550
362	0·002762	501·2	2·700
373	0·002681	760·0	2·881

(a) Plot a graph of $\log_{10} p$ against $1/T$.

(b) Use the graph to obtain a value for ΔH_{vap} of water.

(c) ΔH_{vap} for hydrogen sulphide is 18·8 kJ mole^{-1}. Explain the difference between the values for water and hydrogen sulphide in terms of bonding.

(d) What is the b.pt. of water when the external pressure is 355·1 mm of Hg. Explain your answer.

14 At 60·2°C, the density of a sample of dinitrogen tetroxide is 2·265 kg m^{-3} (2·265 g l^{-1}) at a pressure of 101300 Nm^{-2} (1 atm). Calculate the percentage of nitrogen dioxide molecules in the sample under these conditions. R = 8·314 J mol^{-1} K^{-1} (0·0821 l atm K^{-1} mol^{-1}).

15 (a) Calculate the mole fraction of ethanol in a solution containing 10 g of ethanol and 10 g of water. What is the mole fraction of water in the same solution? (A_r C = 12, H = 1, O = 16).

(b) If the mole fraction of ethanol in a solution of ethanol and water is 0·5, what is the percentage by mass of ethanol?

16 Using the molecular kinetic theory explain the following:

(i) A substance is less compressible in the solid than in the liquid state.

(ii) A solid usually expands slightly on melting.

(iii) A volatile liquid cools when a stream of an inert gas is bubbled through it.

3 Thermochemistry

First law of thermodynamics or law of conservation of energy

This can be stated in 3 ways:

1 Energy can be neither created nor destroyed.
2 The total energy of an isolated system remains constant. An isolated system is one which can neither receive energy from, nor give energy to, anything outside the system.
3 Energy can be converted from one form into an equivalent amount of one or more other forms.

(The law ignores the interconversion of mass and energy according to the relationship $E = mc^2$.)

Energy changes in chemical reactions

Chemical reactions involve an internal energy change given the symbol ΔU. If energy is lost by the system then, by the first law of thermodynamics, this energy must be gained by the surroundings and vice versa.

The internal energy change may involve

(a) heat evolved or absorbed (q) and/or
(b) work done by or on the system (w),

$$\therefore\ \Delta U = q + w.$$

Work done by a system includes

(a) the production of electricity as occurs in cells,
(b) expansion against the atmosphere if there is an increase in the number of gaseous molecules during the reaction, e.g.

$$Zn(s) + 2H^+(aq) \longrightarrow Zn^{2+}(aq) + H_2(g).$$

Work done on a system involves contraction if the reaction leads to a decrease in the number of gaseous molecules.

Following the IUPAC convention all energy **evolved by** the system is negative and all energy **received by** the system is positive.

Work done on expansion or contraction $= \text{force} \times \text{distance moved}$

$$= \frac{\text{force}}{\text{area}} \times \text{distance moved} \times \text{area}$$

$$= \text{pressure} \times \text{volume change}$$

$$= p \times \Delta V.$$

But

$$pV = nRT$$

$$p\Delta V = \Delta nRT.$$

Where Δn is the change in the number of gaseous molecules during the reaction,

ΔnRT is negative for expansion,

ΔnRT is positive for contraction,

$$\therefore \Delta U = q + \Delta nRT.$$

Heat change at constant volume

If a reaction is carried out at constant volume as in the bomb calorimeter then no expansion or contraction can occur.

$$\text{Heat change at constant volume} = q_v = \Delta U.$$

Heat change at constant pressure (enthalpy change ΔH)

The heat change at constant pressure is measured when an experiment is performed in an open vessel such as a calorimeter. The heat exchanged under these conditions is known as the enthalpy change and is given the symbol ΔH.

Then $\Delta U = \text{heat change at constant pressure} + w$

or $\Delta U = \Delta H + \Delta nRT$

or $\Delta H = \Delta U - \Delta nRT.$

(For reactions involving solids and liquids only ΔV is very small and so ΔH approximately equals ΔU.)

The numerical value of an enthalpy change ΔH depends on:

(a) temperature,

(b) the physical states of the reactants or products,

(c) the pressure of any gaseous reactants or products,

(d) the concentrations of any solutions involved.

A standard enthalpy change $\Delta H^\ominus$ is therefore defined. It is the enthalpy change at 298°K for substances in their normal physical states at that temperature. Any gases involved are at a pressure of 1 atmosphere and any solutions are at unit concentration.

For example calculate ΔU for the reaction between zinc and hydrochloric acid if $\Delta H^{\ominus}_{298}$ for the same reaction is -152 kJ.

$$Zn(s) + 2H^+(aq) \longrightarrow Zn^{2+}(aq) + H_2(g)$$

Expansion occurs during the reaction and $\Delta n = 1$.

Work done against the atmosphere $= -\Delta nRT$

$$= -1 \times 8{\cdot}3 \times 298 \text{ J}$$
$$= -2{\cdot}5 \text{ kJ.}$$

Since

$$\Delta U = \Delta H + \Delta nRT$$
$$\Delta U = -152 - 2{\cdot}5$$
$$= -154{\cdot}5 \text{ kJ}$$

Summary

(a) If expansion occurs then more heat is evolved at constant volume than at constant pressure.

(b) If contraction occurs then less heat is evolved at constant volume than at constant pressure.

Enthalpy of reaction

The enthalpy of reaction is the enthalpy change when the number of moles of reactants as indicated in the balanced equation react together, e.g.

$$N_2(g) + 3H_2(g) \longrightarrow 2NH_3(g) \qquad \Delta H^{\ominus}_{298} = -92 \text{ kJ}$$

i.e. when 28 g of nitrogen react with 6 g of hydrogen to form 34 g of ammonia under standard conditions then 92 kJ of heat are evolved at constant pressure.

Enthalpy of formation

The standard enthalpy of formation $\Delta H^{\ominus}_{f,\,298}$ is the enthalpy change when 1 mole of a compound is made from its elements under the standard conditions. The elements must be in their most stable allotropic form at these conditions, e.g. the most stable allotrope of sulphur at 298 K is rhombic sulphur and of carbon is graphite.

$$H_2(g) + \tfrac{1}{2}O_2(g) \longrightarrow H_2O(l)$$
$$\Delta H^{\ominus}_{f,\,298}[H_2O(l)] = -285{\cdot}9 \text{ kJ mol}^{-1}$$

Note: (a) The equation may contain fractional molecules as only 1 MOLE of the compound is formed.

(b) The enthalpy of formation of an element in its standard state is 0, e.g.

$$Hg(l) \longrightarrow Hg(l) \qquad \Delta H^{\ominus}_{f,\,298}[Hg(l)] = 0$$

but

$$Hg(l) \longrightarrow Hg(g) \qquad \Delta H^{\ominus}_{f,\,298}[Hg(g)] = +60{\cdot}8 \text{ kJ mol}^{-1}$$

Enthalpy of combustion

The standard enthalpy of combustion $\Delta H^{\ominus}_{c,\,298}$ is the enthalpy change when 1 mole of a substance undergoes complete combustion (measured under standard conditions), e.g.

$$C_2H_6(g) + \tfrac{7}{2}O_2(g) \longrightarrow 2CO_2(g) + 3H_2O(l) \qquad \Delta H^{\ominus}_{c,\,298}[C_2H_6(g)] = -1559{\cdot}5 \text{ kJ mol}^{-1}$$

Note: (a) ΔH_c is always exothermic.

(b) If the substance contains carbon then carbon dioxide not carbon monoxide is formed.

(c) 1 mole is burnt so the equation may contain fractions.

Enthalpy of atomization of an element

The standard enthalpy of atomization of an element $\Delta H^{\ominus}_{at,\,298}$ is the enthalpy change when 1 mole (g-atom) of gaseous atoms is formed from the element in its physical state under standard conditions, e.g.

$$\tfrac{1}{2}H_2(g) \longrightarrow H(g) \qquad \Delta H_{at,\,298}[\tfrac{1}{2}H_2(g)] = +218 \text{ kJ mol}^{-1}$$

This is half of the energy required to break the single covalent bond between the hydrogen atoms in the molecule, i.e. it is half of the H—H bond energy. $\Delta H^{\ominus}_{at,\,298}$ is always endothermic.

Enthalpy of atomization of a compound

The standard enthalpy of atomization of a compound $\Delta H^{\ominus}_{at,\,298}$ is the enthalpy change when 1 mole of the compound is converted into gaseous atoms under standard conditions, e.g.

$$CH_4(g) \longrightarrow C(g) + 4H(g) \qquad \Delta H^{\ominus}_{at,\,298}[CH_4(g)] = +1662 \text{ kJ mol}^{-1}$$

Enthalpy of neutralization

The standard enthalpy of neutralization of an acid by a base is the enthalpy change when the mass of acid capable of producing 1 mole of $H^+(aq)$ ions in solution is neutralized by the equivalent amount of the base under standard conditions, the reaction being carried out in dilute aqueous solution, e.g.

$$\mathrm{NaOH(aq) + HCl(aq) \longrightarrow NaCl(aq) + H_2O(l)} \qquad \Delta H = -57{\cdot}3\ \mathrm{kJ}$$

Since NaOH and HCl are examples of strong alkalis and acids they are completely dissociated in solution and the essential reaction is

$$\mathrm{OH^-(aq) + H^+(aq) \longrightarrow H_2O(l)} \qquad \Delta H = 57{\cdot}3\ \mathrm{kJ}$$

Thus the enthalpy of neutralization of any strong acid by any strong base is always $-57{\cdot}3$ kJ.

However, if a weak acid, e.g. ethanoic acid, or a weak base, e.g. ammonia solution, is used then dissociation is incomplete and less heat will be evolved.

Hess's law of constant heat summation

This law states that the total energy change accompanying a chemical reaction is independent of the route by which the change takes place, i.e. if,

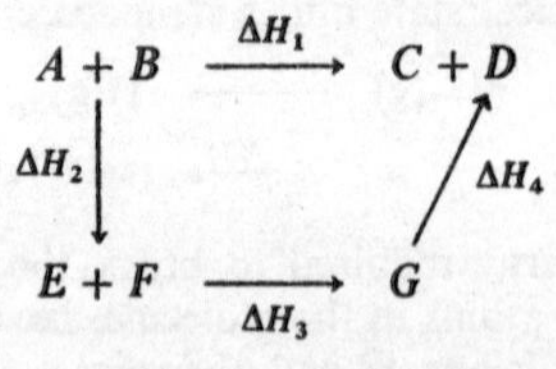

then $\Delta H_1 = \Delta H_2 + \Delta H_3 + \Delta H_4$

Hess's law follows from the first law of thermodynamics. Its use lies in the calculation of ΔH values which are difficult to obtain experimentally.

(a) Calculate $\Delta H^\ominus_{f,\,298}$ of H_2S given:

$$\begin{aligned} \Delta H^\ominus_{c,\,298}\quad \mathrm{H_2(g)} &= -285{\cdot}9 \\ \mathrm{S(g)} &= -296{\cdot}6 \\ \mathrm{H_2S(g)} &= -561{\cdot}4\ \mathrm{kJ\ mol^{-1}} \end{aligned}$$

Equation required:

$$\mathrm{H_2(g) + S(s) \longrightarrow H_2S(g)} \qquad \Delta H^\ominus_{f,\,298} = ?$$

Given:

$$H_2(g) + \tfrac{1}{2}O_2(g) \longrightarrow H_2O(l) \qquad ①$$
$$\Delta H^{\ominus}_{c,\,298} = -285{\cdot}9\ \text{kJ}$$

$$S(s) + O_2(g) \longrightarrow SO_2(g) \qquad ②$$
$$\Delta H^{\ominus}_{c,\,298} = -296{\cdot}6\ \text{kJ}$$

$$H_2S(g) + \tfrac{3}{2}O_2(g) \longrightarrow H_2O(l) + SO_2(g) \qquad ③$$
$$\Delta H^{\ominus}_{c,\,298} = -561{\cdot}5\ \text{kJ}$$

Add ① and ② and subtract ③ to obtain the required equation.

$$\therefore\ \Delta H^{\ominus}_{f,\,298}[H_2S(g)] = -285{\cdot}9 + (-296{\cdot}6) - (-561{\cdot}4)\ \text{kJ}$$
$$= -21{\cdot}1\ \text{kJ}$$

This may be represented diagrammatically

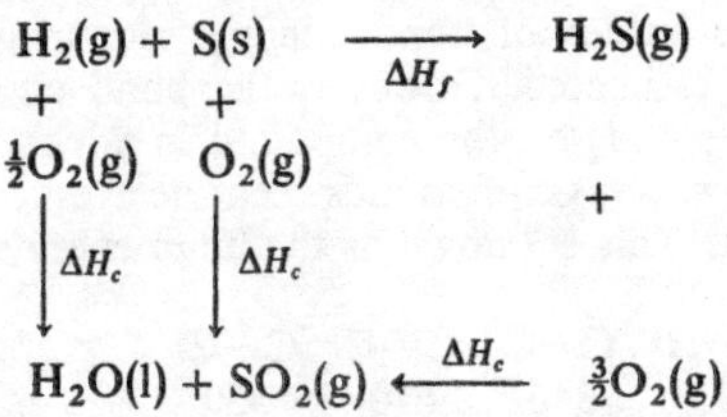

By Hess's law

$$\Delta H_f[H_2S(g)] = \Delta H_c[H_2(g)] + \Delta H_c[S(s)] - \Delta H_c[H_2S(g)]$$

(b) Calculate $\Delta H^{\ominus}_{c,\,298}[CH_4(g)]$ given:

$$\Delta H^{\ominus}_{f,\,298}\quad CH_4(g) = -75$$
$$O_2(g) = 0$$
$$CO_2(g) = -393{\cdot}5$$
$$H_2O(l) = -285{\cdot}9\ \text{kJ mol}^{-1}$$

Equation required:

$$CH_4(g) + 2O_2(g) \longrightarrow CO_2(g) + 2H_2O(l)$$
$$\Delta H^{\ominus}_{c,\,298} = ?$$

Given:

$$C(gr) + 2H_2(g) \longrightarrow CH_4(g) \qquad ①$$
$$\Delta H^{\ominus}_{f,\,298} = -75\ \text{kJ}$$

$$O_2(g) \longrightarrow O_2(g) \qquad ②$$
$$\Delta H^{\ominus}_{f,\,298} = 0$$

$$C(gr) + O_2(g) \longrightarrow CO_2(g) \qquad ③$$
$$\Delta H^{\ominus}_{f,\,298} = -393{\cdot}5\ \text{kJ}$$

$$H_2(g) + \tfrac{1}{2}O_2(g) \longrightarrow H_2O(l) \qquad ④$$
$$\Delta H^{\ominus}_{f,\,298} = -285{\cdot}9\ \text{kJ}$$

Add ③ and ② × ④, subtract ① and ② × ②.

$$\therefore \Delta H^{\ominus}_{c,298}[CH_4(g)] = -393{\cdot}5 + (2 \times -285{\cdot}9) - (-75 + 2 \times 0)\ \text{kJ mol}^{-1}$$
$$= -890{\cdot}3\ \text{kJ mol}^{-1}$$

Note: For all ΔH_c calculations the following relationship applies

$$\Delta H_c = \sum \Delta H_f \text{ of products} - \sum \Delta H_f \text{ of reactants.}$$

Bond energy terms

The enthalpy change in a chemical process = $\sum \Delta H$ bonds formed $- \sum \Delta H$ bonds broken.

Thus if accurate values of bond energies are available enthalpies of reaction can be estimated. However, the bond energy of a particular covalent bond depends on the compound in which it is situated. If its value is found in various compounds then the average per mole of bonds can be calculated. This is known as the bond energy term $\bar{E}$, e.g.

Bond	C—H	C—C	C=C	C—Cl	C—Br	C—I	N—H	O—H	S—H
$\bar{E}$ kJ mol^{-1}	413	346	610	339	284	218	391	463	344

Values of $\bar{E}$ can be used in estimating enthalpies of reaction.

For example, the reduction of an alkene to an alkane involves the following bond changes:

		$\bar{E}$			$\bar{E}$
Bonds broken	C=C	610	Bonds made	C—C	346
	H—H	433		2(C—H)	2 × 413
	total	1043		total	1172

Theoretically the reaction therefore involves the evolution of 129 kJ. This agrees with the values obtained from experiments.

The sum of the bond energy terms in a particular compound approximately equals the enthalpy of atomization of that compound.

Evaluation of the C—H *bond energy term*

The energy of the C—H bond is found in various ways. For example:

(a) Use of the enthalpy of atomization of methane. From experiment

$$CH_4(g) \longrightarrow C(g) + 4H(g)$$
$$\Delta H^{\ominus}_{at,\,298} = +1662\ \text{kJ mol}^{-1}$$

∴ The bond energy of the C—H bond in methane

$$= \frac{1662}{4}$$

$$= 415{\cdot}5 \text{ kJ mol}^{-1}.$$

(b) Use of the enthalpies of atomization of butane and pentane.

$$C_4H_{10}(g) \longrightarrow 4C(g) + 10H(g)$$

$$\Delta H^{\ominus}_{at,\,298} = +5165 \text{ kJ mol}^{-1}$$

$$C_5H_{12}(l) \longrightarrow 5C(g) + 12H(g)$$

$$\Delta H^{\ominus}_{at,\,298} = +6337 \text{ kJ mol}^{-1}$$

To atomize butane 2 C—C bonds and 10 C—H bonds are broken.
To atomize pentane 4 C—C bonds and 12 C—H bonds are broken.

∴ $3E\,(\text{C—C}) + 10E\,(\text{C—H}) = +5156$

and $4E\,(\text{C—C}) + 12E\,(\text{C—H}) = +6337$

Solving the equations

$E(\text{C—C}) = +347 \text{ kJ mol}^{-1}$

and $E(\text{C—H}) = +412 \text{ kJ mol}^{-1}$

Considering these and other values, the C—H bond energy term is 413 kJ mol^{-1}.

Measurement of enthalpy changes

1 **Calorimetric methods** for reactions in solutions, e.g. precipitation, neutralization etc.

Method

(a) The temperatures of the reactants *A* and *B* (known volumes of known molarities) are recorded every minute for about 5 minutes.

(b) The reactants are mixed in a lagged polystyrene cup and the temperature recorded every 30 seconds for about 10 minutes.

(c) A graph of temperature against time is plotted.

The graph is extrapolated to the time of mixing to allow for cooling and any delay in temperature registration.
Rise in temperature of reactant $A = t_1$°C.
Rise in temperature of reactant $B = t_2$°C.
Assuming the specific heat capacity of water and dilute solutions is 4·18 J g^{-1} K^{-1} and that the heat capacity of the calorimeter and stirrer is 0.

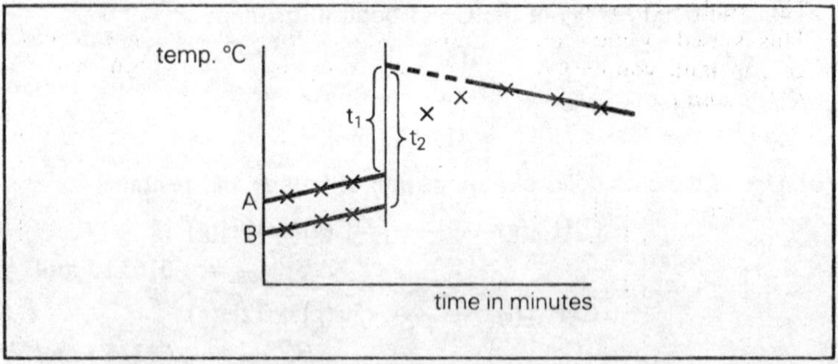

Then heat change = $(m_A \times 4{\cdot}18 \times t_1) + (m_B \times 4{\cdot}18 \times t_2)$ J where m_A and m_B are the masses of reactants A and B.
This is the enthalpy change for forming x moles of the product. Thus the enthalpy change for 1 mole is found.

2 Combustion calorimeter

Apparatus similar to that shown is used in finding the enthalpy of combustion of inflammable liquids.

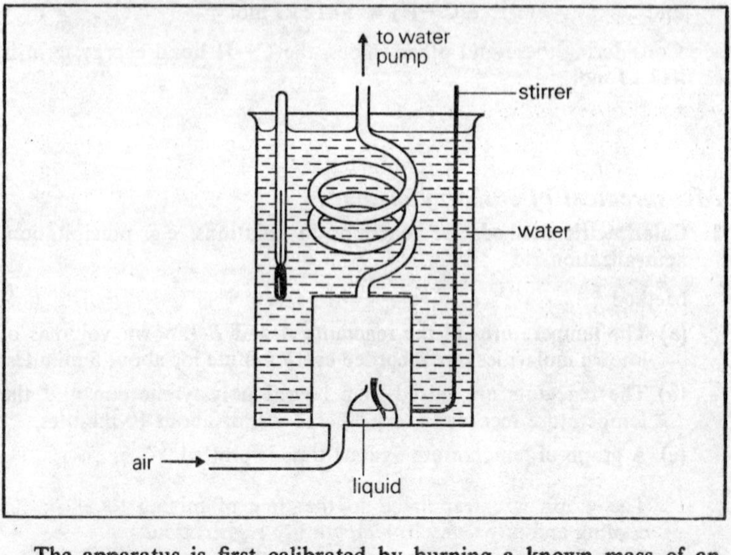

The apparatus is first calibrated by burning a known mass of an alcohol of known enthalpy of combustion. From the temperature rise the heat capacity of the apparatus is determined. The experiment is repeated with a known mass of the liquid of unknown enthalpy of combustion.

3 The bomb calorimeter

This is used to find ΔU_c of solids or liquids since the reaction occurs at constant volume. ΔH_c is found using the relationship $\Delta H = \Delta U - \Delta nRT$.

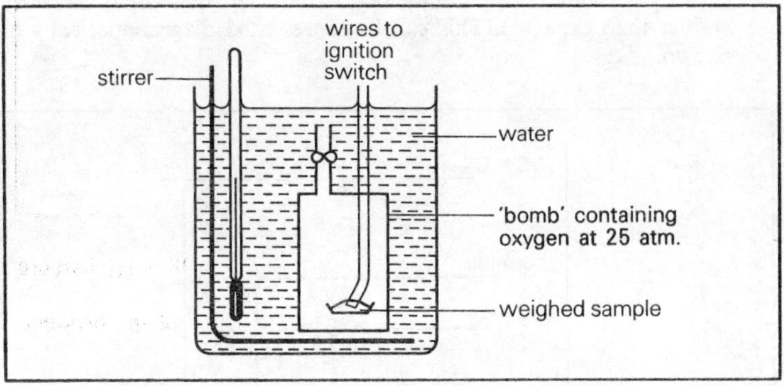

The known mass of sample in the platinum crucible is ignited by passage of a current through the wire. The temperature rise is noted. The apparatus is calibrated by burning a known mass of a substance of known ΔH_c.

Delocalization energy

Thermochemical data indicates that certain compounds containing multiple covalent bonds are stabler than expected. This may be explained by assuming that the π electrons in the multiple bonds are delocalized over the whole or part of the species. The extra stability is then known as the delocalization energy, e.g. benzene.

(a) Enthalpy of hydrogenation

A value for the enthalpy of hydrogenation of a C=C is obtained by the hydrogenation of cyclohexene.

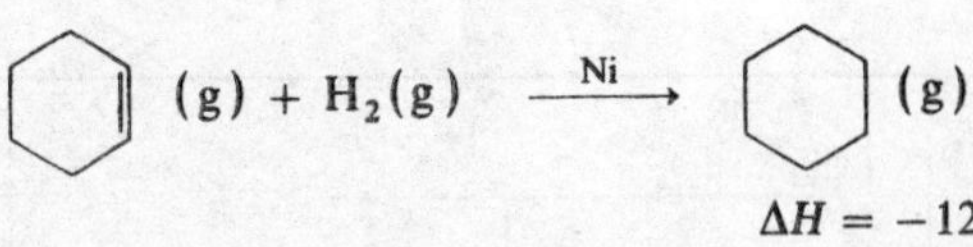

$\Delta H = -120 \text{ kJ mol}^{-1}$

If benzene is regarded as containing 3 isolated C=C then the expected enthalpy of hydrogenation in the gaseous phase is -360 kJ mol^{-1}, i.e.

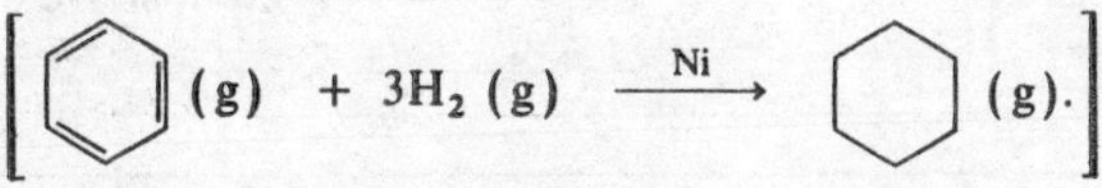

However, the experimentally determined value is only -208 kJ mol^{-1}. This gives a value of 152 kJ mol^{-1} for the delocalization energy of benzene. To explain this, it is assumed that the π electrons in the 3 double bonds are delocalized over the whole of the molecule, giving each C—C bond partial double bond character. Benzene is therefore stabler than expected. This can be represented diagrammatically as shown.

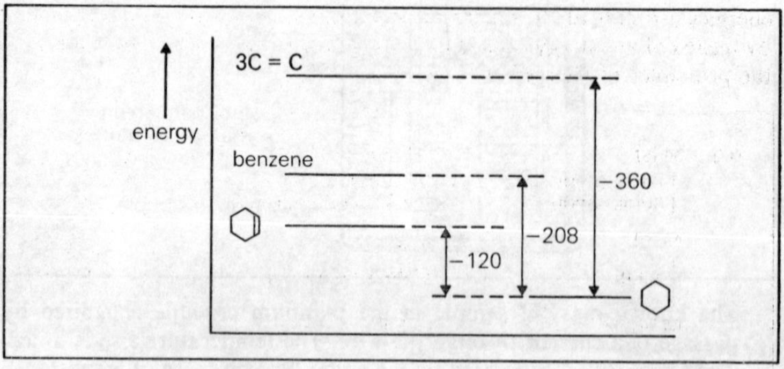

(b) **Enthalpies of formation of gaseous benzene**
The enthalpy of formation of gaseous benzene is calculated from bond energy terms assuming 3 isolated C=C.

$$\begin{aligned}\Delta H_f &= -[6\bar{E}(\text{C—H}) + 3\bar{E}(\text{C—C}) + 3\bar{E}(\text{C=C})] \\ &= -[(6 \times 413) + (3 \times 346) + (3 \times 610)]\ \text{kJ mol}^{-1} \\ &= -5346\ \text{kJ mol}^{-1}.\end{aligned}$$

However, the value obtained from experimental ΔH combustion measurements = -5514 kJ mol^{-1}.
∴ Delocalization energy = 168 kJ mol^{-1}.

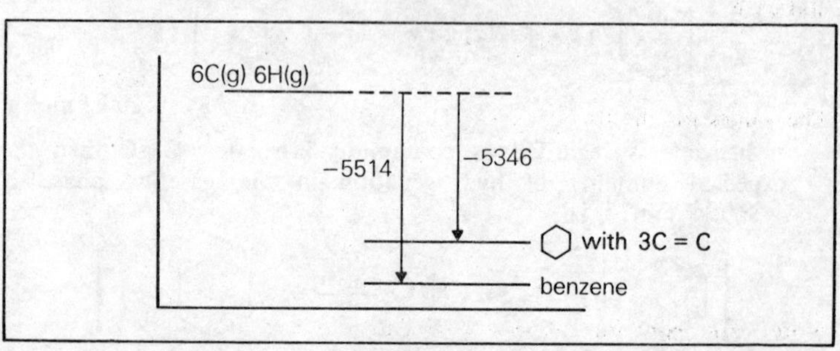

Lattice energy

The lattice energy is the enthalpy change when 1 mole of an ionic compound is formed from its constituent ions in the gaseous state, e.g.

$$Na^+(g) + Cl^-(g) \longrightarrow NaCl(s)$$

$$\Delta H^{\ominus}_{298} = \text{lattice energy}$$

Its value gives a measure of the sizes of the inter-ionic forces. Lattice energies are difficult to determine experimentally so they are calculated by means of an energy cycle called the Born-Haber cycle. This embodies the principles of Hess's law.

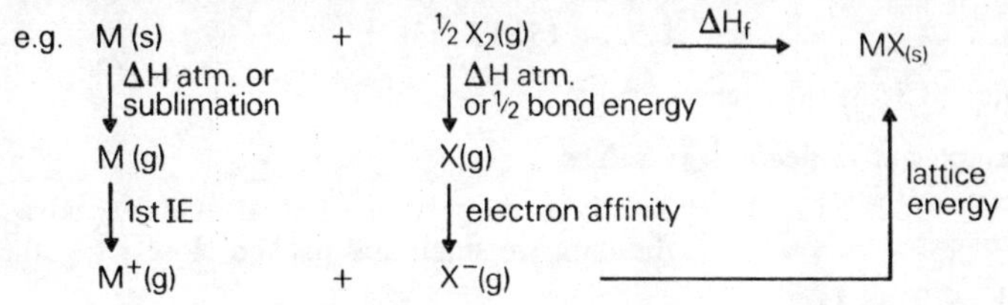

$$\Delta H_f = \Delta H_{atm.} + IE + \Delta H_{atm.} + EA + \text{lattice energy}$$

For example for NaCl,

$\Delta H_{at}[Na(s)] = +108{\cdot}3\ kJ\ mol^{-1}$, 1st *IE* Na $= +500\ kJ\ mol^{-1}$.

$\Delta H_{at}[Cl(g)] = +121{\cdot}1\ kJ\ mol^{-1}$, *EA* Cl $= -364\ kJ\ mol^{-1}$.

$\Delta H_f[NaCl] = -411\ kJ\ mol^{-1}$.

$\therefore$ Lattice energy $= -411 - 108{\cdot}3 - 500 - 121{\cdot}1 + 364\ kJ\ mol^{-1}$

$= -776{\cdot}4\ kJ\ mol^{-1}$.

The **electron affinity** is the enthalpy change when 1 mole of atoms in the gaseous state each accept 1 electron to form 1 mole of ions in the gaseous state, e.g.

$$Cl(g) + e \longrightarrow Cl^-(g)$$

$$\Delta H = -364\ kJ\ mol^{-1}$$

The values for the halogens are:

F	$-332{\cdot}6\ kJ\ mol^{-1}$
Cl	$-364\ kJ\ mol^{-1}$
Br	$-342\ kJ\ mol^{-1}$
I	$-295{\cdot}4\ kJ\ mol^{-1}$

Note: the value for chlorine is largest.

The second electron affinity is usually positive because of the repulsive effect of the negative charge, e.g.

$$O(g) + e \longrightarrow O^-(g) \qquad \Delta H = -141{\cdot}4 \text{ kJ mol}^{-1}$$

$$O^-(g) + e \longrightarrow O^{2-}(g) \qquad \Delta H = +790{\cdot}8 \text{ kJ mol}^{-1}$$

Bond energy terms for the halogens are:

F_2	+ 158 kJ mol^{-1}
Cl_2	+ 242 kJ mol^{-1}
Br_2	+ 193 kJ mol^{-1}
I_2	+ 151 kJ mol^{-1}

Note: Cl_2 has the largest value.

Examples of lattice energy values

LiF	− 1029 kJ mol^{-1}	This is the largest for all alkali metal halides as the ions are small and packed closely together.
NaF	− 915 kJ mol^{-1}	
CsI	− 594 kJ mol^{-1}	This is the smallest value for alkali metal halides as the ions are large.
MgO	− 3933 kJ mol^{-1}	The value is large as the ions are doubly charged.

Theoretical values for lattice energies

The lattice energy is the energy evolved when well-separated electrostatic charges are brought together to form a lattice. This energy can be calculated if each ion is regarded as a spherical, separate entity with its charge distributed uniformly over it. The total lattice energy is composed of 2 parts.

(a) **Attractive forces**

The attractive force between 2 ions of charge $+e$ and $-e$, a distance r apart $= e^2/4\pi\varepsilon r^2$, ε is the permittivity of a vacuum.

The energy released on bringing them together from infinity to the interionic distance d in the crystal

$$= \int_{\infty}^{d} \frac{e^2}{4\pi\varepsilon r^2} = \frac{e^2}{4\pi\varepsilon r}.$$

Similarly the energy involved in bringing up successive ions to form a row and then a 2 and 3 dimensional lattice is calculated.

(b) **Repulsive forces**

The ions repel each other when they are so close together that their negative electronic clouds surrounding their nuclei interact. Repulsive forces fall off rapidly with distance.

The calculated lattice energy values can be compared with those obtained by applying the Born-Haber cycle to experimentally measured enthalpies.

Compound	Theoretical value kJ mol^{-1}	Experimental value kJ mol^{-1}
NaCl	−766·1	−776·4
NaI	−685·7	−688·3
KCl	−692·0	−697·8
KI	−630·9	−631·8
AgCl	−768·6	−916·3
AgI	−735·9	−865·4
ZnS	−3427	−3615

For alkali metal halides the 2 values are very similar indicating that the ionic model is satisfactory.

However, with the silver and zinc salts there is considerable discrepancy indicating that a degree of covalency is present.

Lattice energy and stoichiometry
Lattice energy can be used to indicate the likely formula of a compound.

Using the Born-Haber cycle with reasonable estimates of the lattice energy values of $MgCl_2$ and the hypothetical compounds MgCl and $MgCl_3$, their enthalpies of formation are calculated.

$$\Delta H^{\ominus}_{f,\,298}[\mathrm{MgCl(s)}] = -113\ \mathrm{kJ\ mol^{-1}}$$
$$\Delta H^{\ominus}_{f,\,298}[\mathrm{MgCl_2(s)}] = -653\ \mathrm{kJ\ mol^{-1}}$$
$$\Delta H^{\ominus}_{f,\,298}[\mathrm{MgCl_3(s)}] = +3904\ \mathrm{kJ\ mol^{-1}}$$

$MgCl_2$ is therefore more energetically stable with respect to its elements than MgCl and $MgCl_3$. This indicates that magnesium chloride will probable have the formula $MgCl_2$.

Practice questions

In questions 1 to 14 give the letter of the correct answer.

1 Under which of the following conditions will the lattice energy of an ionic crystal *AB* have the largest value?
A. Both A^+ and B^- have small radii.
B. *A* has a small ionization energy.
C. *B* has a small electron affinity.
D. *A* has a large sublimation energy.
E. A^+ has a small radius, B^- has a large radius.

2 In which of the following processes is the associated enthalpy change negative?

A. $K(s) \longrightarrow K(g)$ B. $Br(g) + e \longrightarrow Br^-(g)$
C. $Br_2(l) \longrightarrow 2Br(g)$ D. $O^-(g) + e \longrightarrow O^{2-}(g)$
E. $NaCl(s) \longrightarrow Na^+(g) + Cl^-(g)$

3 For which of the following is the associated enthalpy change equal to the lattice energy of MgO?

A. $Mg(s) + \frac{1}{2}O_2(g) \longrightarrow MgO(s)$
B. $Mg(g) + O(g) \longrightarrow MgO(s)$
C. $Mg^{2+}(g) + \frac{1}{2}O_2(g) \longrightarrow MgO(s)$
D. $Mg^{2+}(g) + O^{2-}(g) \longrightarrow MgO(s)$
E. $Mg(s) + O(g) \longrightarrow MgO(s)$

4 Which of the following describes the relationship between the heat change at constant volume (ΔU) and that at constant pressure (ΔH) for the reaction:

$$2C_2H_2(g) + 5O_2(g) \longrightarrow 4CO_2(g) + 2H_2O(g)$$

A. $\Delta H = \Delta U$.
B. Both ΔH and ΔU are positive.
C. ΔH is positive and ΔU is negative.
D. ΔH is negative and ΔU is positive.
E. ΔU has a more negative value than ΔH.

5 Which of the following is the value of $\Delta H^{\ominus}_{c,\,298}[CH_3OH(l)]$ if $\Delta H^{\ominus}_{f,\,298}$ of $CO_2(g)$, $H_2O(l)$ and $CH_3OH(l)$ are $-393{\cdot}5$, $-285{\cdot}9$ and $-238{\cdot}9$ kJ mol^{-1} respectively?

A. $-918{\cdot}3$ kJ mol^{-1} B. $-1204{\cdot}2$ kJ mol^{-1}
C. $-726{\cdot}4$ kJ mol^{-1} D. $+1204{\cdot}2$ kJ mol^{-1}
E. $-440{\cdot}5$ kJ mol^{-1}

Questions 6 to 8

Choose from the list A to E, the enthalpy change associated with the reactions in questions 6 to 8 (heat evolved is given a negative sign).

A. +2 kJ B. +240 kJ C. −59 kJ D. −3 kJ E. −2800 kJ

6 $KOH(aq) + HNO_3(aq) \longrightarrow H_2O(l) + KNO_3(aq)$
7 C (diamond) $\longrightarrow$ C (graphite)
8 $C_4H_8(g) + 6O_2(g) \longrightarrow 4CO_2(g) + 4H_2O(g)$

Questions 9 to 14

Choose from the list A to E the correct description of the enthalpy change associated with the conversions in questions 9 to 14.

A. Enthalpy of formation
B. Enthalpy of ionization
C. Enthalpy of atomization
D. Bond energy
E. Enthalpy of neutralization

9 $C(gr) + 2H_2(g) \longrightarrow CH_4(g)$

10 $I(g) \longrightarrow I^+(g) + e$

11 $HS^-(aq) + OH^-(aq) \longrightarrow H_2O(l) + S^{2-}(aq)$

12 $H_2(g) \longrightarrow 2H$

13 $\frac{1}{2}I_2(g) \longrightarrow I(g)$

14 (a) Define enthalpy of combustion of a compound and state the extra conditions which must be imposed to specify the standard enthalpy of combustion.

(b) Give the equation for the complete combustion of ethanol. Calculate $\Delta H_c^\ominus$ for ethanol given:

$\Delta H_f^\ominus$ for ethanol (l) $= -277{\cdot}0$ kJ mol^{-1}
$\Delta H_f^\ominus$ for water (l) $= -285{\cdot}9$ kJ mol^{-1}
$\Delta H_f^\ominus$ for carbon dioxide (g) $= -393{\cdot}7$ kJ mol^{-1}

15 The graph shows the temperature change when 1 M NaOH is added to 25 cm^3 of HCl in a polystyrene cup. The solutions were initially at the same temperature.

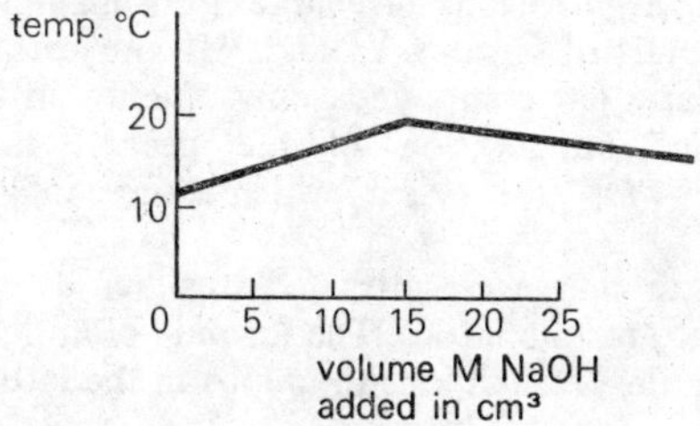

(a) Use the graph to
  (i) find the molarity of the HCl,
  (ii) calculate ΔH neutralization of NaOH.
  (Assume the specific heat capacity of all solutions is 4·2 kJ K^{-1}.)

(b) Give the value of ΔH neutralization you would expect for 1 mole of KOH reacting with 1 mole CH_3COOH.

4 Bonding and Crystal Structure

1 *Ionic bonding*

Ionic bonding involves the transfer of 1 or more electrons from a metal to a non metal with the formation of ions.

Metals forming positive ions have low ionization energies and the ion formed usually has the electronic configuration of the nearest noble gas, e.g. Mg ($1s^22s^22p^63s^2$) $\rightarrow 2e + Mg^{2+}$ ($1s^22s^22p^6$), which is the configuration of Ne. However this is not necessarily true for transition metal ions. Thus Cr^{3+} has the configuration $1s^22s^22p^63s^23p^63d^3$. Cations with a charge of greater than 3+ are rarely found as the energy required in their formation is large and the ion produced would have a large charge to size ratio making it highly polarizing. Most metals form ionic compounds although certain compounds of Li, Be, Al and some transition metals have a large degree of covalency.

Non-metals forming negative ions have large electron affinity values. They include the non-metals of Groups VI and VII and certain of those in Groups IV and V in a few compounds only. The anion formed usually has the electronic configuration of the nearest noble gas, e.g. O ($1s^22s^22p^4$) $+ 2e \rightarrow O^{2-}$ ($1s^22s^22p^6$) which is the configuration of Ne.

The ions of opposite charge are attracted together by non-directional electrostatic forces to form a lattice. The formula of an ionic compound is the ratio of the number of cations to anions in the lattice.

Properties of ionic compounds

1 Ionic compounds are solids with high melting points.

2 Ionic compounds are usually soluble in polar but not in non-polar solvents. On dissolution two energy processes are involved:
   (a) The lattice energy is required to separate the ions.
   (b) The enthalpies of solution of the ions are evolved. If the solvent is water the term hydration is used. The water molecule is polar

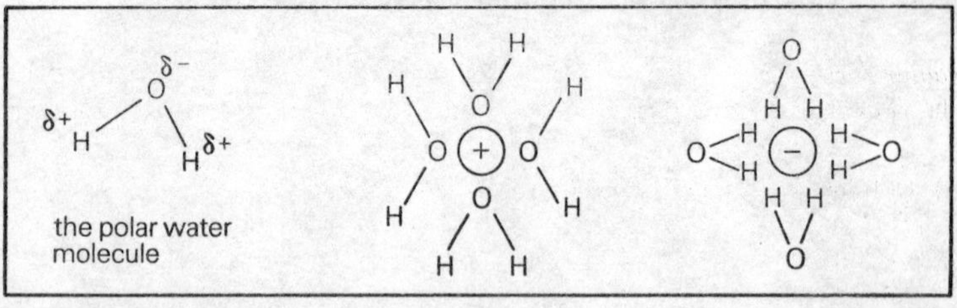

in nature as oxygen is more electronegative than hydrogen and because the molecule is non-linear. The water molecules stabilize the free ions by forming solvent shells as shown.

The enthalpy of hydration helps to counteract the lattice energy. However, some compounds have very high lattice energies and so are insoluble, e.g. $Ba^{2+}SO_4^{2-}$. The enthalpy of solution at infinite dilution is the enthalpy change when 1 mole of a compound dissolves in such a large volume of solvent that the addition of more solvent produces no further heat change. It is the difference between the energy changes (a) and (b) above and so may be positive or negative. An example of a compound dissolving exothermically is anhydrous copper(II) sulphate and endothermically is ammonium nitrate. (To explain why certain compounds with endothermic enthalpies of solution do dissolve, free energy and entropy factors must also be considered.)

3 Solid ionic compounds are non-conductors of electricity but do conduct when molten or in solution. The ions are then free to move and carry the current.

Lattice structures

The arrangement of the ions in the crystal depends mainly on the relative sizes of the cations and anions concerned (the radius ratio).

Ionic radii

Values for ionic radii are found by measuring the interionic distances in crystals using X-ray diffraction techniques. The graph on p. 66 shows the variation of atomic and ionic radii with atomic number.

(a) Atomic radii and also to some extent ionic radii vary periodically with *Z*.

(b) Anionic radii are larger than the atomic radii of the atoms concerned. Cationic radii are smaller than the atomic radii of the atoms concerned, e.g. Cl 0·099 nm, Cl^- 0·181 nm, Na 0·157 nm, Na^+ 0·098 nm.

(c) Both atomic and ionic radii increase down a group, e.g.

F	0·072	Cl	0·099	Br	0·114	I	0·133 nm
F^-	0·136	Cl^-	0·181	Br^-	0·195	I^-	0·216 nm

Radius ratio

The ratio is the ratio of the cationic to anionic radii.

(a) **Cations and anions of the same size, i.e. radius ratio = 1.**
The most stable arrangement is one in which each cation is surrounded by 6 anions in the same plane and 3 above and 3 below. The co-ordination number of the cation is therefore 12.

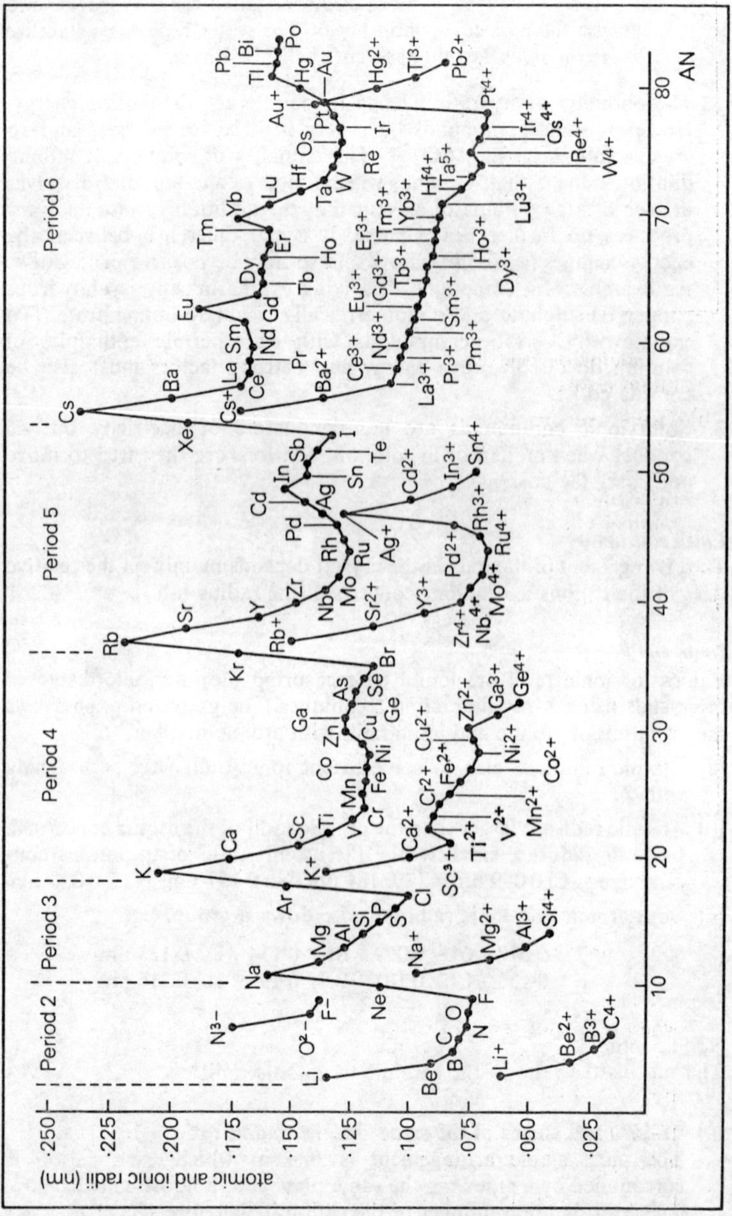
atomic and ionic radii (nm)
·250
·225
·200
·175
·150
·125
·100
·075
·050
·025
Period 2
Period 3
Period 4
Period 5
Period 6
10
20
30
40
50
60
70
80
AN

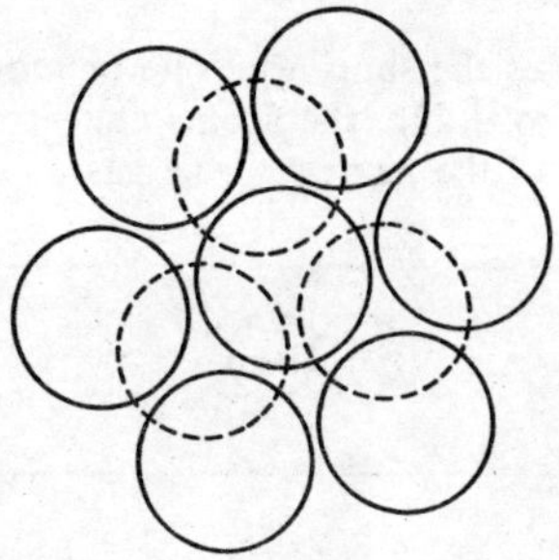

To maintain electrical neutrality, the co-ordination number of the anion is also 12.

(b) **Cations smaller than anions, i.e. radius ratio is less than 1.**
The co-ordination number is less than 12. The diagram represents the smallest cation capable of being surrounded by 4 anions in the same plane and one above and one below, i.e. the co-ordination number is 6.

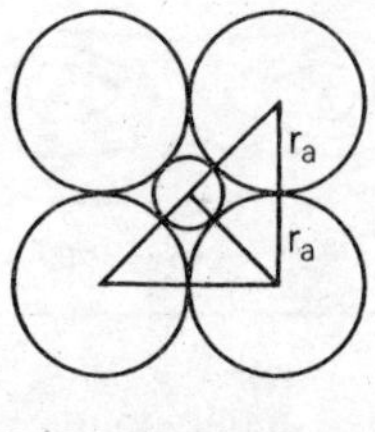

$$r_c = \sqrt{2}\, r_a - r_a$$

$$\frac{r_c}{r_a} = \sqrt{2} - 1$$

$$= 0{\cdot}414$$

where r_a = anionic radius
r_c = cationic radius

If $r_c/r_a < 0{\cdot}414$ the co-ordination number is less than 6.

If r_c/r_a has a value between 0·414 and 0·732, the co-ordination number is 6, e.g. NaCl, NaBr, KCl, KBr.

If $r_c/r_a > 0{\cdot}732$, the co-ordination number is 8, e.g. CsCl, CsBr, CsI.

2 *Covalent bonding*

Covalent bonding involves the sharing of one or more pairs of electrons between two atoms. In so doing, the atoms concerned often obtain the electronic configuration of the nearest noble gas.

Each hydrogen atom in the molecule has the configuration of He. This type of representation is a means of counting electrons but does not indicate the positions of the electrons which are distributed in space as diffuse negative charge clouds.

The following diagram represents an electron density map for the hydrogen molecule.

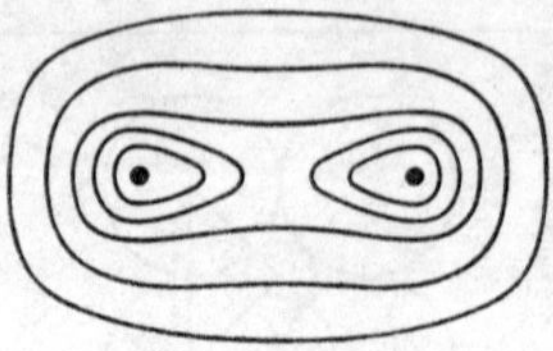

There is an electron density at all points along the line joining the 2 nuclei, i.e. the electrons are shared. In structures containing ions, the electron density drops to zero between the ions.

Further examples of covalent compounds

1 Methane CH_4

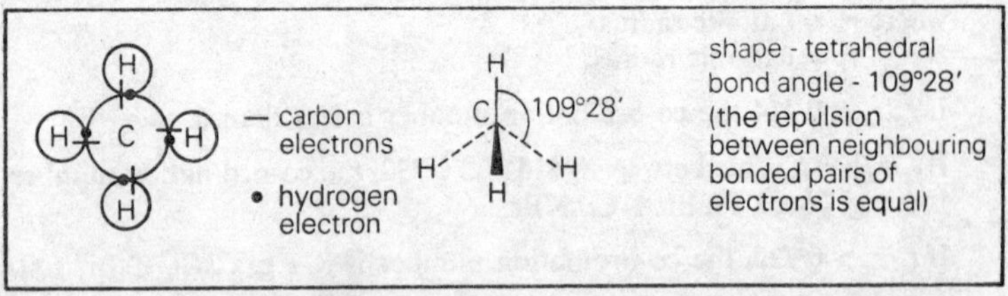

2 Ammonia NH_3

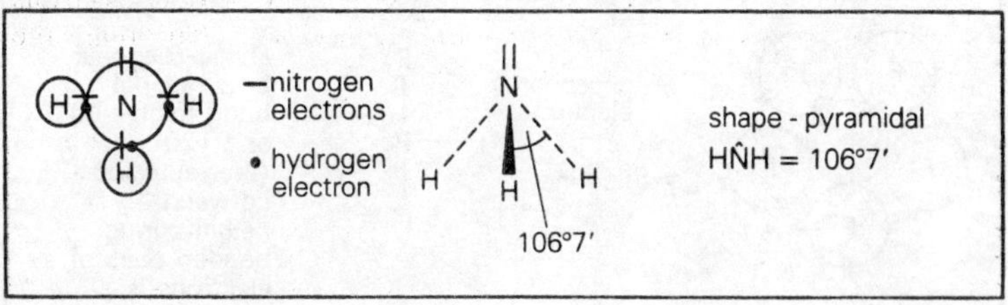

The lone pair–bond pair repulsion is greater than the repulsion between 2 bonded pairs of electrons so the HN̂H is slightly less than the tetrahedral value.

3 Water H_2O

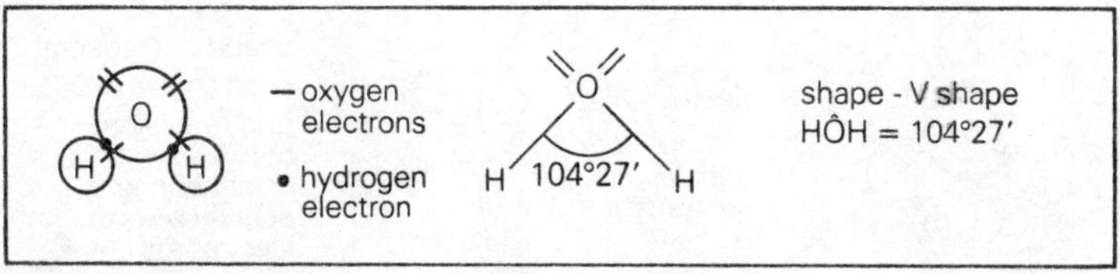

The lone pair–lone pair repulsion > lone pair–bond pair repulsion > bond pair–bond pair repulsion so the HÔH is smaller than the tetrahedral angle.

4 Boron trifluoride BF_3

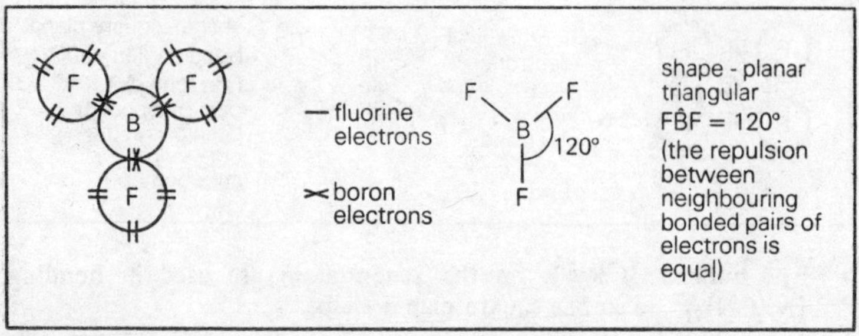

5 Phosphorus pentafluoride PF_5

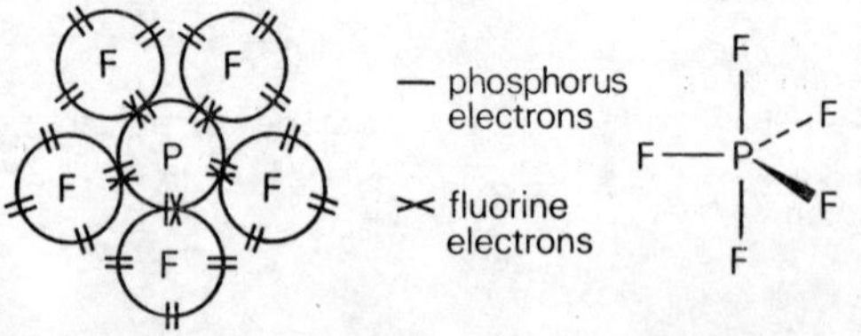

shape-triagonal bi-pyramidal
bond angles 90° and 120°
(the repulsion between neighbouring bonded pairs of electrons is equal)

The unfilled 3*d* orbitals on the phosphorus atom are used in the bonding. $Fe(CO)_5$ is also trigonal bipyramidal.

6 Sulphur hexafluoride SF_6

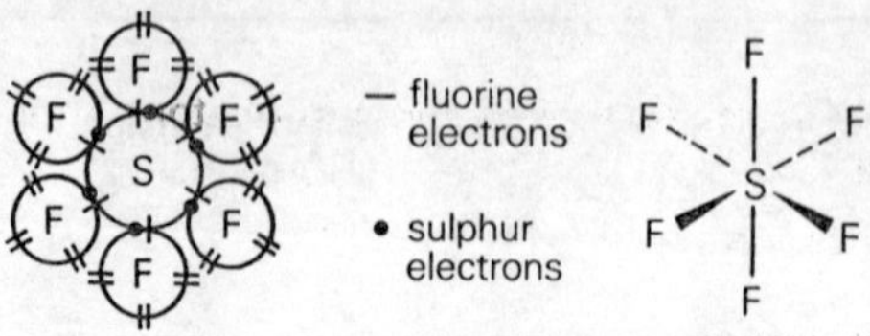

shape - octahedral
bond angles - 90°
(the repulsion between neighbouring bonded pairs of electrons is equal)

The unfilled 3*d* orbitals on the sulphur atom are used in bonding. Transition metal ion complexes containing 6 ligands are usually regular octahedral in shape, e.g. $[Fe(H_2O)_6]^{3+}$.

7 Xenon tetrafluoride XeF_4

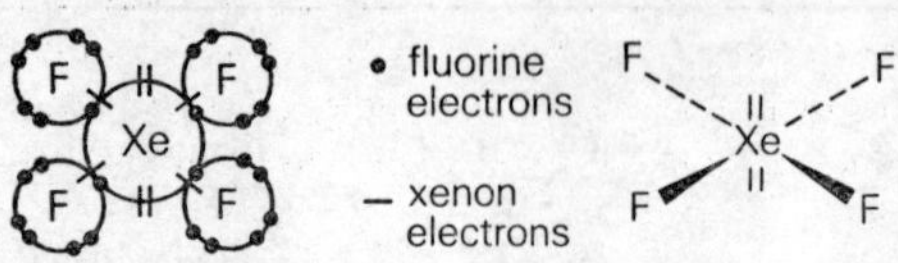

shape - square planar
bond angles - 90°
(the lone pairs occupy opposite positions in the octahedral arrangement)

The unfilled 4*f* levels on the xenon atom are used in bonding. $[Ni(CN)_4]^{2-}$ also has square planar shape.

8 Ethene C_2H_4

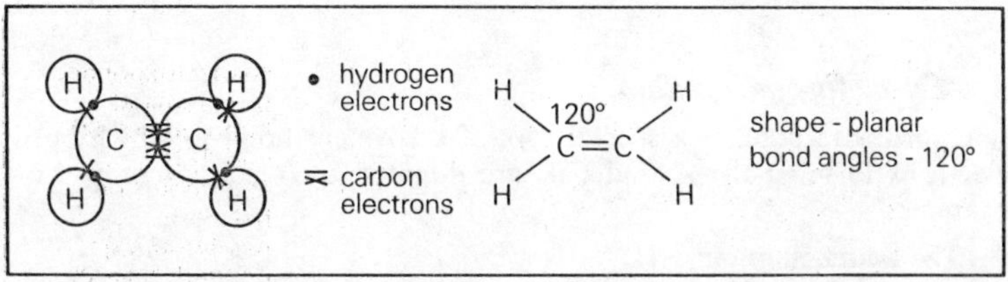

Two pairs of electrons are shared between the carbon atoms forming a double covalent bond. The molecule is therefore unsaturated.

9 Ethyne C_2H_2

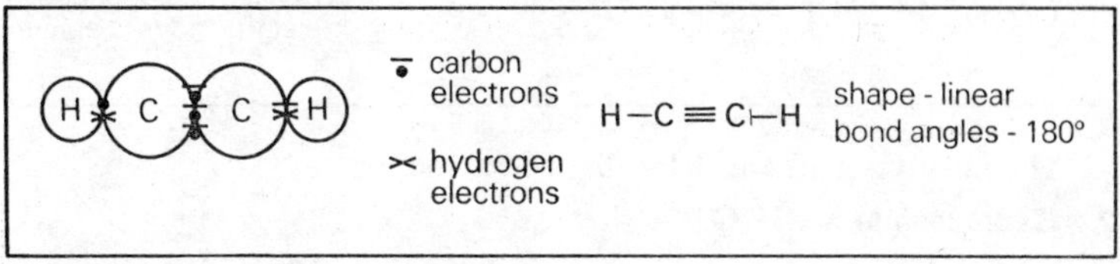

Three pairs of electrons are shared between the carbon atoms forming a triple covalent bond.

Properties of covalent compounds

1 At room temperature covalent compounds are usually gases, liquids or low melting point solids. The forces between the molecules are weak although the covalent bond itself is strong.

2 Covalent compounds are soluble in non-polar solvents such as benzene unless they interact in some way with polar solvents.

3 Covalent compounds when solid, melted or in solution do not conduct.

Atomic and covalent radii

The atomic radius of a particular element is half the distance between the nuclei of two adjacent atoms in the close packed structure or molecule of the element. For elements forming covalent molecules the term 'covalent radius' is used. The graph on page 66 shows the variation of atomic radius with atomic number.

Bond length

The bond length of a covalent bond is the distance between the nuclei in the bond. It is found by X-ray or spectroscopic methods and is usually independent of the nature of the other groups attached to the bond.

In general, a reasonable estimate of any bond length is found by adding together the appropriate covalent radii.

3 *Co-ordinate bonding*

A co-ordinate bond is a special case of a covalent bond in which both electrons forming the shared pair are donated by 1 species only. For example:

1 The ammonium ion $NH_4{}^+$

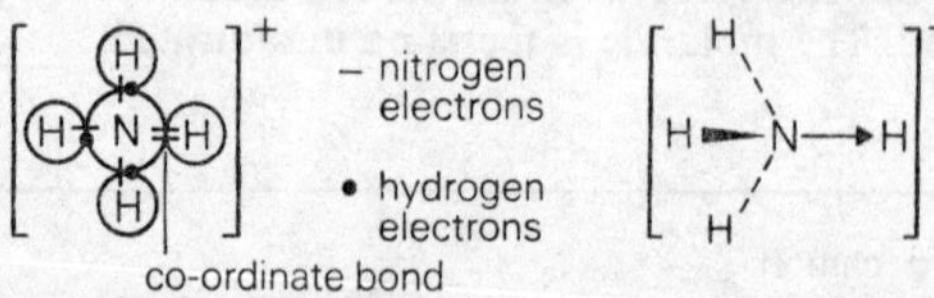

shape: tetrahedral
bond angles: 109° 28′
(the repulsion between the bonded electron pairs is equal)

The positive charge is delocalized over the ion.

2 Hydroxonium ion H_3O^+

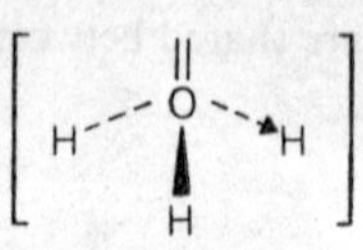

3 Transition metal ion complexes

In many of these complexes the ligands are the electron-pair donors and form co-ordinate bonds with the transition metal ion.

4 Dimers such as Al_2Cl_6, Fe_2Cl_6 and chains such as $(BeCl_2)_n$

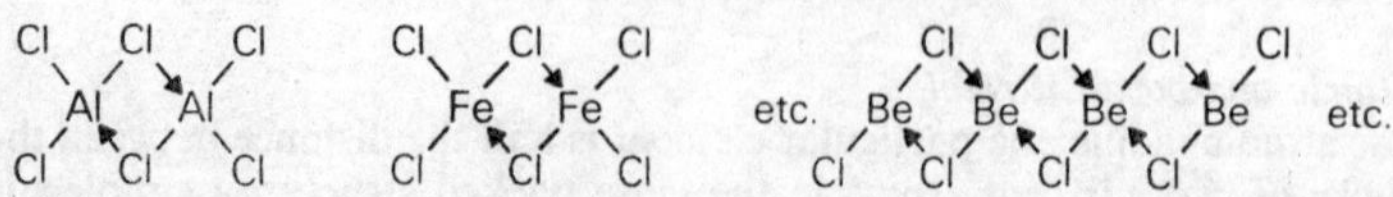

5 Ammonia boron trifluoride adduct and ammonia boron trichloride adduct

6 Delocalization of electrons in covalent molecules and ions

In some molecules and ions, the bond lengths found experimentally do not agree with those calculated from covalent radii. For example:

(a) Benzene C_6H_6

A possible structural formula for benzene could be

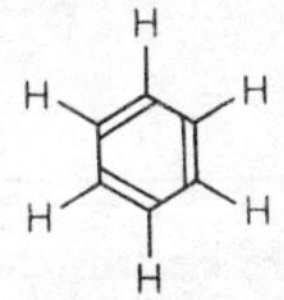

Single bond covalent radius of C = 0·077 nm.
Double bond covalent radius of C = 0·067 nm.
∴ Calculated C—C bond length in benzene = 0·154 nm.
Calculated C=C bond length in benzene = 0·134 nm.
However X-ray and electron diffraction experiments show the molecule is a regular, planar hexagon with C—C bond lengths of 0·140 nm. This is explained by assuming that the 1 electron in the p_z orbital on each C atom is not localized between 2 C atoms in forming a double bond but is delocalized over the ring. Thus each C—C bond has partial double bond character. Further evidence for the existence of delocalization is provided by thermochemical data (see previous chapter).

(b) Ethanoate, nitrate and sulphate anions

Possible structural formulae

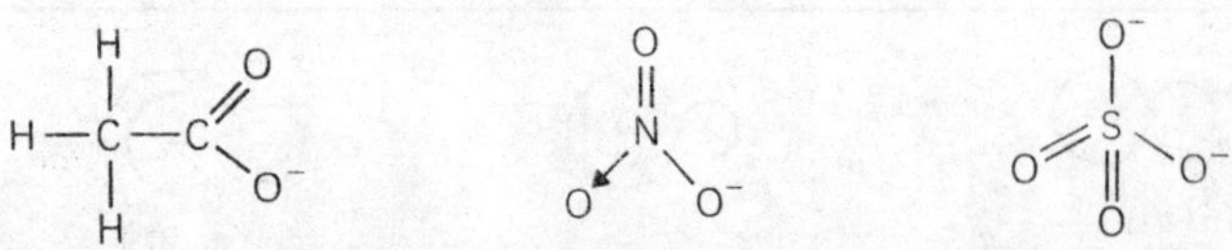

The above structural formulae suggest that in each anion there would be 2 X—O bond lengths corresponding to the different types of bonding present. However, in all cases it is found that the X—O bond lengths in the particular ion are the same and somewhere between the 2 expected values. This is best explained by assuming that the π electrons in the multiple bonds are delocalized over the whole of the ion.

Nature of the covalent bond

In covalent molecules there is a high electron density in the region between the nuclei. For example consider the H_2^+ ion (1 electron system).

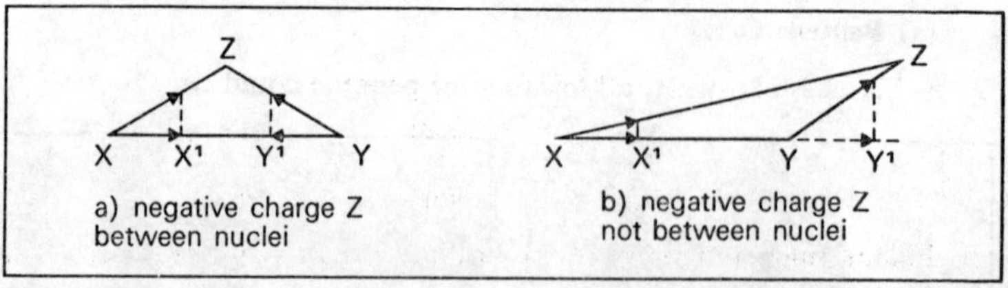

Diagrams (a) and (b) show the electrostatic forces on the nuclei caused by a small portion of the negative charge Z.

When it is between the nuclei (diagram (a)) it attracts them both equally causing them to be drawn towards each other (resolved parts XX^1 and YY^1). When it is not between the nuclei (diagram (b)) it attracts Y more than X thus tending to separate them.

However, since there is more negative charge cloud between the nuclei than outside them there is a net attractive force binding the atoms together.

Polarization

A particular compound is said to be polarized if it shows both ionic and covalent characteristics.

(a) **Polarization of ions**

In many ionic compounds the ions are not spherical but are distorted or polarized. The electron density between the ions is therefore increased and a degree of covalency results.

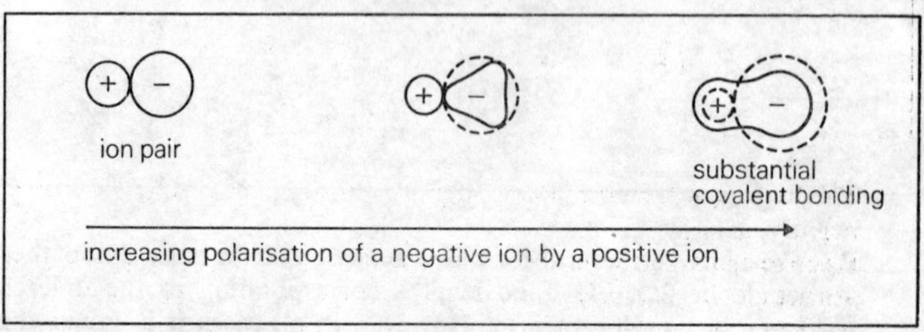

Evidence for a degree of covalent bonding in ionic compounds is obtained from:

(i) Lattice energy measurements (see chapter 3).

(ii) Spectroscopic measurements. The distances between the nuclei in certain alkali metal halides are found to be smaller in the vapour than in the solid phases. This implies the existence of stronger bonding and therefore a degree of covalency in the vapour.

Internuclear separation in nm

	Crystal	**Vapour**
LiBr	0·275	0·217
LiI	0·300	0·239

Fajans' rules summarize the conditions which favour polarization of negative ions:

(i) A small positive ion.
(ii) A high charge on the positive ion.
(iii) A large negative ion (the electron cloud is more easily distorted as it is less strongly held).
(iv) A large charge on the negative ion.

(b) **Polarization of covalent bonds**

If the electron pair in a covalent bond is not shared equally between the 2 atoms then the bond is said to be polarized and the molecule containing the bond may have a permanent electrostatic dipole moment.

Dipole moments are vector quantities and the value obtained for a polyatomic molecule is a resultant of the individual bond dipoles. The resultant is zero in linear molecules of the type *X*-*Y*-*X*, e.g. CO_2 and in tetrahedral molecules of the type YX_4, e.g. CCl_4. The degree of polarization depends on the relative electronegativities of the atoms concerned. For example in the hydrogen chloride molecule, the electron charge cloud is distorted towards the chlorine atom as chlorine is the more electronegative.

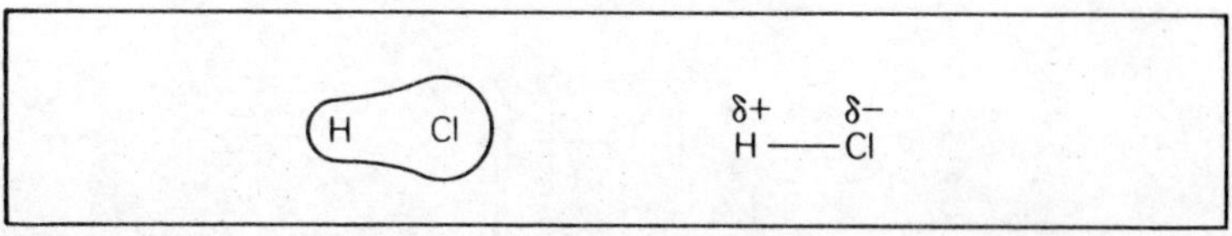

The hydrogen therefore has a slight positive and the chlorine a slight negative charge.

Electronegativity is the ability of an atom in a covalent bond to attract electrons and so produce an inductive effect. The value for a particular atom depends on its environment and so may vary from compound to compound. (Electron affinity is concerned with isolated atoms.) Values of electronegativities have been calculated from bond energy terms (Pauling's values) and from ionization energies and electron affinities (Mulliken's values).

The trends in electronegativity values in the Periodic Table are as indicated below.

increasing electronegativity →

Li	Be	B	C	N	O	F	↑ increasing electronegativity
Na						Cl	
K						Br	
Rb						I	
Cs							

F is the most electronegative and Cs the least. On the Pauling scale they have the values 4·0 and 0·7 respectively.
In general, the larger the difference in electronegativity between the atoms forming the bond, the greater the ionic character, e.g. BeF_2 is more ionic than $BeCl_2$.

4 *Metallic bonding*

In solid metals the atoms are packed closely together in regular array in either close-packed or body-centred cubic systems.

(a) There are two close-packed structures:
   (i) Hexagonal close packing, e.g. Mg, Zn, Ni. Each atom is surrounded by 6 neighbours in the plane of the layer and every alternate layer lies directly on top of each other, i.e. it is of the type *abab* etc.
   (ii) Face-centred cubic close-packing, e.g. Cu, Ag, Au, Al. Each atom is surrounded by 6 neighbours in the plane of the layer and third layer is directly on top of each other, i.e. it is of the type *abcabc* etc.

(b) Body-centred cubic, e.g. Na, K. This is not close-packed but contains about one third empty space.

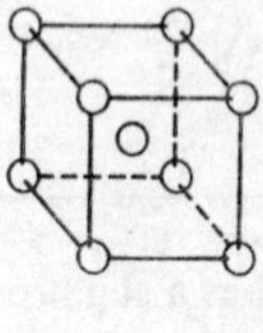

The metallic bond

The conducting property of metals is best explained by the assumption that each atom donates some of its electrons to a delocalized electron cloud. The positive metal ions are embedded in the charge cloud and held in position by electrostatic forces. When an electrical potential difference is applied, the electron plasma flows.

Properties

(a) Metals are good conductors of heat and electricity. The increased oscillation of an ion brought about by heat is transmitted to neighbouring ions. This impedes the electron flow so the electrical conductivity decreases with temperature.

(b) Pure metals are malleable and ductile as the spherical ions can 'slide' over each other. This effect is reduced by the addition of impurities.

Intermolecular forces

1 **Electrostatic interaction** between molecules with dipole moments (see chapter 2).

2 **Van der Waals' forces** between non-polar molecules or atoms. Slight displacement of the nuclei or electrons gives rise to transitory electrical dipoles in molecules such as O_2, N_2, CCl_4 etc. and noble gas atoms. The displacement is temporary and random so over time the molecule or atom has no permanent dipole. In any given instant the dipole will induce another dipole in a neighbouring molecule or atom and they will be attracted together. These forces are very weak and are called van der Waals' forces. They are more significant in solids and liquids when the particles are close together.

3 **Hydrogen bonding** between molecules containing hydrogen linked to very electronegative atoms such as fluorine, nitrogen or oxygen. The electronegative atom attracts the electron pair in the covalent bond and acquires a slight negative charge. This leaves the hydrogen atom with an equal slight positive charge. The hydrogen is then attracted to the more negative part of the neighbouring molecule, e.g. in hydrogen fluoride vapour, $(HF)_5$ molecules are found to exist. The hydrogen bond is electrostatic in nature.

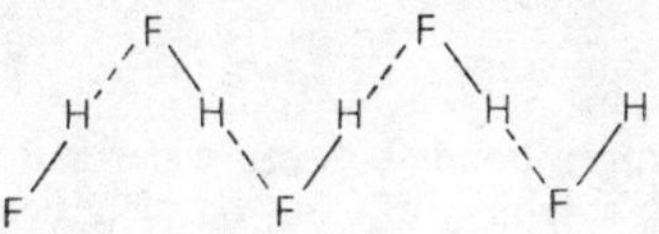

--- denotes hydrogen bonding

Properties of hydrogen bonds

(a) Hydrogen bonds, like covalent bonds, are directional in nature.

(b) Hydrogen bonds are stronger than van der Waals' forces but much weaker than the forces involved in ionic and covalent bonds.

Hydrogen bond	Approximate energy kJ mol^{-1}	Covalent bond	Bond energy kJ mol^{-1}
—H---F	125	H—F	554
—H---O	25	H—O	464
—H---N	20	H—N	389

(c) Hydrogen bonding influences physical but not chemical properties. For example:

(i) Melting points and boiling points
The melting and boiling points of H_2O, HF and NH_3 are higher than expected from M_r considerations.

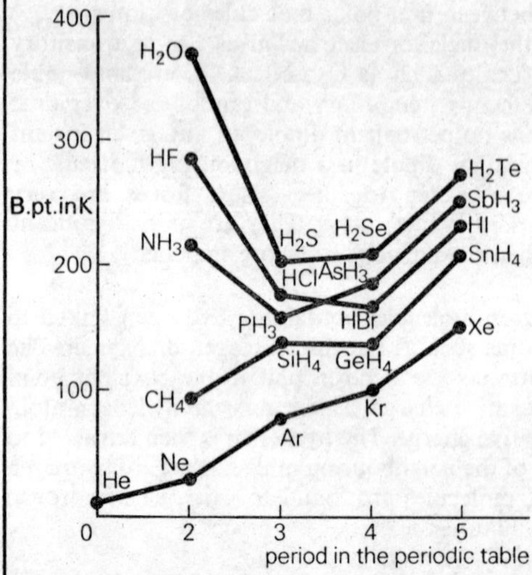

water forms on average 2 hydrogen bonds per molecule at room temperature. In ice there are 4 per molecule and an open 3 dimensional lattice is formed which is essentially tetrahedral. Water therefore expands on freezing and the ice formed has a low density

(ii) Solubility
The formation of hydrogen bonds between water and solute is responsible for the high solubility of many covalent compounds such as sugar, ammonia and hydrogen fluoride.

(iii) $\Delta H^{\ominus}_{fusion}$ and $\Delta H^{\ominus}_{vap}$
The molar latent heats of fusion and vaporization of compounds containing hydrogen bonding are higher than expected from M_r considerations, e.g.

	$\Delta H^{\ominus}_{fusion}$ kJ mol^{-1}	$\Delta H^{\ominus}_{vap}$ kJ mol^{-1}
NH_3	5·65	23·4
PH_3	1·13	14·6
H_2O	6·02	40·7
H_2S	2·39	18·7

Examples of compounds and ions containing hydrogen bonding

1 $[HF_2]^-$ ion
This is linear with equal HF bond lengths, $[F\text{---}H\text{---}F]^-$.

2 Carboxylic acids in gaseous and liquid phases. Dimers are formed, e.g.

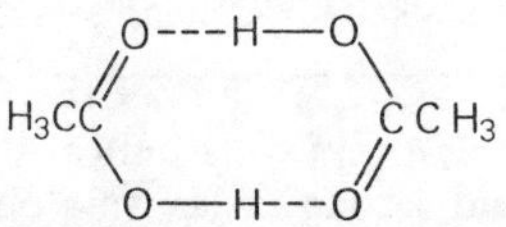

3 Solid hydrates, e.g. gypsum $CaSO_4.2H_2O$.
Gypsum consists of layers of Ca^{2+} and $SO_4{}^{2-}$ ions linked together by hydrogen bonds between the water molecules and $SO_4{}^{2-}$ ions. The crystals are easily cleaved as the hydrogen bonds are weak.

4 Structures in living systems.
The 3 dimensional structures of e.g. proteins, enzymes, DNA are maintained by hydrogen bonds.

Crystal structures

X-Ray diffraction

X-ray diffraction provides information on the arrangement of the particles in a crystal. When a beam of X-rays is directed onto a crystal face, the beam is diffracted by the layers of particles. These diffracted rays can be regarded as though they are reflected.

Thus, for X-rays of a given wavelength directed onto a certain face of a crystal, maximum reinforcement of the diffracted rays occurs for fixed values of glancing angles θ. These are found experimentally by varying the glancing angle and recording its value when strong signals are obtained in the detector. The distance between the layers of particles parallel to the face is calculated. The experiment is repeated using the other faces of the crystal.

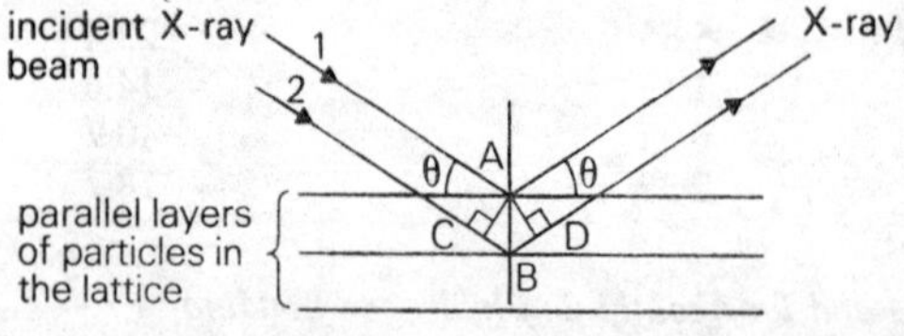

If the reflected rays are to reinforce each other (to be in phase) then the extra distance travelled by the wave 2 must equal a whole number of wavelengths.

i.e. $BC + BD = n\lambda$

$BC = BD = AB \sin\theta$ where θ is the angle of incidence or glancing angle

$\therefore n\lambda = 2AB \sin\theta$

If the distance between parallel layers of particles in the crystal is d then $n\lambda = 2d \sin\theta$. This is known as Bragg's equation.

The unit cell is the simplest repeating unit of the crystal. For example reflections of the first and second order were obtained from 1 face of a sodium chloride crystal at glancing angles of 5·90° and 11·85° respectively using X-rays of wavelength 58·5 pm (1 picometre = 10^{-12} m). Calculate the distance between successive planes of ions parallel to the face.

Using $n\lambda = 2d \sin\theta$

When $n = 1$, $\theta = 5{\cdot}90$ $\sin\theta = 0{\cdot}103$ $\therefore d = \dfrac{58{\cdot}5}{2 \times 0{\cdot}103} = 284{\cdot}0$

When $n = 2$, $\theta = 11{\cdot}85$ $\sin\theta = 0{\cdot}205$ $\therefore d = \dfrac{2 \times 58{\cdot}5}{2 \times 0{\cdot}205} = 285{\cdot}4$

The average distance between successive planes of ions

= 285 pm (corr. to 3 sig. figs.).

Ionic crystal structures

The radius ratio is important in determining the type of crystal structure (see ionic bonding).

1 Sodium chloride

The sodium chloride structure consists of two inter-penetrating face-centred cubic lattices, one of chloride and one of sodium ions.

Number of ions per unit cell.

	Na^+	Cl^-
At the corners, 8 ions, shared by 8 cells	$\frac{1}{8} \times 8$	
On the edges, 12 ions, shared by 4 cells		$\frac{1}{4} \times 12$
On the faces, 6 ions, shared by 2 cells	$\frac{1}{2} \times 6$	
Inside the cell, 1 ion.		1
	4	4

Unit cell

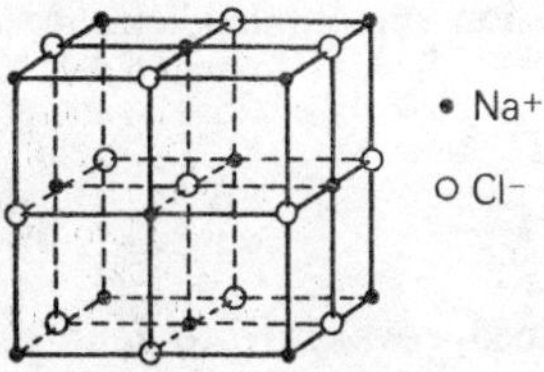

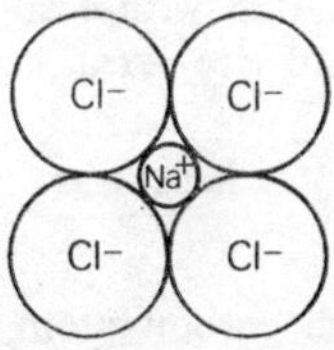

One layer of sodium chloride crystal. In the second layer each sodium ion lies directly above a chloride ion and vice versa. The third layer is the same as the first.

The unit cell contains the equivalent of 4 Na^+ and 4 Cl^- ions giving an empirical formula of NaCl. The co-ordination number is 6 : 6.

2 Caesium chloride

The caesium chloride structure consists of two inter-penetrating simple cubic lattices, one of caesium and one of chloride ions.

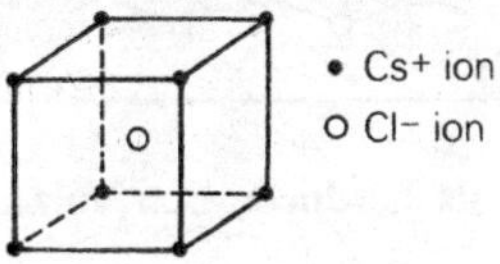

Unit cell

number of ions per unit cell

	Cs^+	Cl^-
at the corners, 8 ions	$\frac{1}{8} \times 8$	
inside the cell, 1 ion		1
Total	1	1

The unit cell contains the equivalent of 1 Cs^+ and 1 Cl^- ion giving the empirical formula CsCl. The co-ordination number is 8 : 8.

3 Calcium fluoride

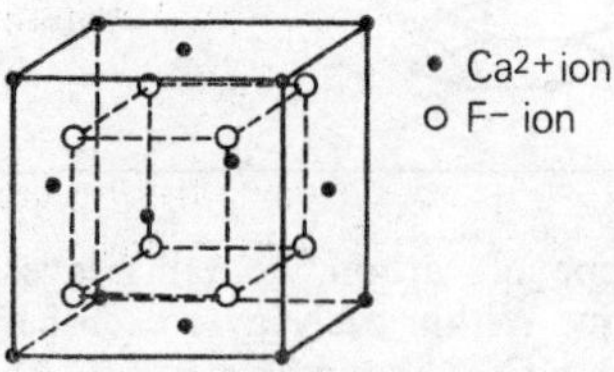

Calcium fluoride has cubic structure. The fluoride ions lie on a simple cubic lattice and the calcium ions on a face-centred cubic lattice. These interpenetrate as shown.

Each Ca^{2+} ion is surrounded by 8 F^- ions at the corners of a cube and each F^- by 4 Ca^{2+} ions arranged tetrahedrally. The co-ordination numbers are 8 and 4.

4 Zinc sulphide (zinc blende structure)

Zinc sulphide has a cubic structure similar to that of diamond. It has 4,4 co-ordination with each Zn^{2+} ion surrounded tetrahedrally by 4 S^{2-} ions and vice versa.

Macromolecular crystal structures

1 Diamond

In diamond each carbon atom is joined covalently to four other carbon atoms arranged tetrahedrally around it. The co-ordination number is 4 and the crystal is cubic.

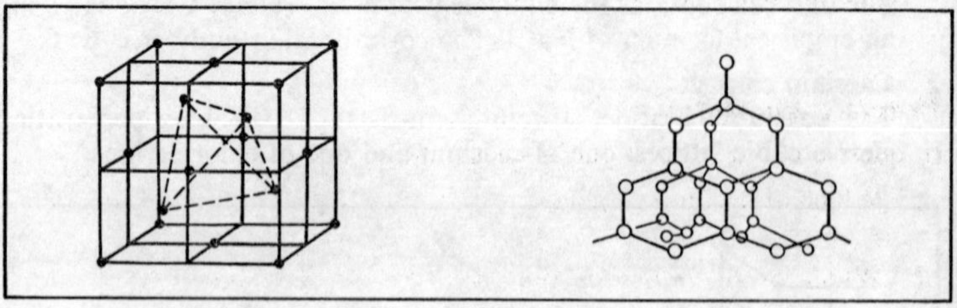

The structure of diamond accounts for its hardness, non volatility, insolubility and unreactivity.

2 Graphite

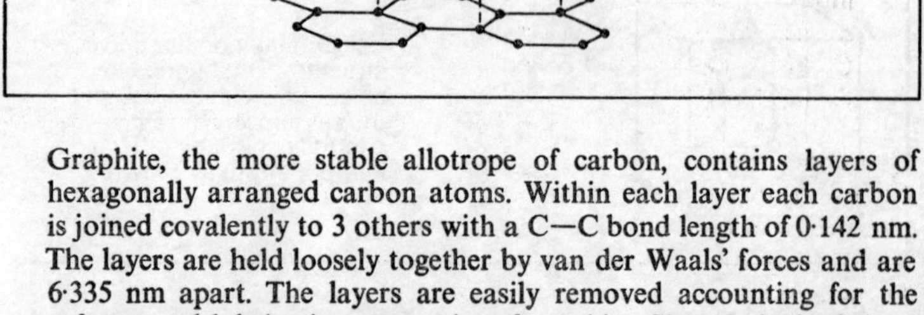

Graphite, the more stable allotrope of carbon, contains layers of hexagonally arranged carbon atoms. Within each layer each carbon is joined covalently to 3 others with a C—C bond length of 0·142 nm. The layers are held loosely together by van der Waals' forces and are 6·335 nm apart. The layers are easily removed accounting for the softness and lubricating properties of graphite. Since only 3 of the 4 outer electrons on each carbon is used in forming covalent bonds, graphite is a good conductor of electricity.

Molecular crystal structures

1 Iodine

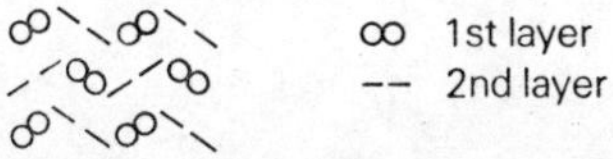

The covalent molecules are arranged in a herringbone pattern as shown. They are held together by weak van der Walls' forces so the solid is volatile (it sublimes) and dissolves in non-polar solvents.

Practice questions

In questions 1 to 18 give the letter of the correct answer.

1 Which of the following has the greatest degree of covalent character?
A. $SrCl_2$ B. $BaCl_2$ C. $CaCl_2$ D. $MgCl_2$ E. $BeCl_2$

2 Which of the following contains at least 1 co-ordinate bond?
A. Fe_2Cl_6 B. NH_3 C. BCl_3 D. $AlCl_3$ E. $SnCl_4$

3 If X has an AN of 5, which of the following is the $F\widehat{X}F$ bond angle in XF_3?
A. 60° B. 90° C. 109° D. 120° E. 180°

4 Which of the following gives the correct bond angle(s) in a molecule of the hydride of an element X, where X has the electronic configuration $1s^22s^22p^3$?
A. 120° B. 72° C. 90 and 120° D. 107·3° E. 109·5°

5 Which of the following has the greatest polarizing power?
A. Li^+ B. Na^+ C. K^+ D. Cs^+ E. I^-

6 Which of the following lists the ions Al^{3+}, Fe^{2+} and Fe^{3+} in decreasing order of polarizing power if the ionic radii are Al^{3+} 0·05 nm, Fe^{2+} 0·076 nm, Fe^{3+} 0·065 nm?
A. Al^{3+}, Fe^{2+}, Fe^{3+} B. Al^{3+}, Fe^{3+}, Fe^{2+}
C. Fe^{2+}, Fe^{3+}, Al^{3+} D. Fe^{2+}, Al^{3+}, Fe^{3+}
E. Fe^{3+}, Al^{3+}, Fe^{2+}

Questions 7 to 12

Choose from the list A to E the correct descriptions of the structures of the species in questions 7 to 12.

A. square planar B. tetrahedral C. trigonal planar
D. pyramidal E. trigonal bipyramidal

7 BCl_3

8 NH_3

9 PF_3

10 PF_5

11 $[Ni(CN)_4]^{2-}$

12 XeF_4

Questions 13 to 18

Choose from the list A to E the conceptual model most useful in explaining the phenomena in questions 13 to 18.

A. Co-ordinate bonding B. Delocalization of electron
C. Hydrogen bonding D. Noble gas configuration
E. Repulsion of electron pairs

13 Silver is a better conductor of electricity than graphite.

14 The S—O bond lengths in the SO_4^{2-} ion are all equal.

15 Ethanoic acid forms dimers under certain conditions.

16 Methane is tetrahedral not planar.

17 Ammonia can act as a Lewis base.

18 The molar latent heat of vaporization of PH_3 is 14·6 and of NH_3 is 23·4 kJ mol^{-1}.

19 Element *A* is in Group I of the Periodic Table and has an electronegativity of 2·7. Element *B* is in Group VII and its electronegativity is 2·7. The electronegativity of hydrogen is 2·1.

(a) Give the formula of the compounds formed between (i) *A* and hydrogen, (ii) *B* and hydrogen. In each case discuss the bonding present.

(b) Explain why the compound in (a)(i) melts with decomposition at about 500°C whilst that in (a)(ii) is a gas at room temperature but dissolves in water to form an acidic solution.

20 Describe how X-rays can be used to investigate the structure of crystals. Outline the structure of 1 ionic, 1 molecular and 1 macromolecular crystal. Indicate how the physical properties of a crystal are dependent on the nature of its lattice.

21 (a) Using the electron repulsion theory, describe and explain the shapes of the following species, giving bond angles.
(i) BF_3 (ii) H_2O (iii) NH_3 (iv) $[PF_6]^-$ (v) NH_2^-
(vi) PCl_4^+ (vii) ICl_4^-

(b) Explain why
(i) BF_3 and PF_5 react with ammonia but not methane,
(ii) the boiling point of water is higher than that of hydrogen sulphide,
(iii) water and ammonia react with anhydrous copper(II) sulphate but methane does not.

22 (a) The following data relates to crystalline sodium chloride; density 2·165 g cm^{-3}, internuclear distance Na^+ to Cl^- 0·2821 nm and and relative formula mass 58·44.

(i) Draw a diagram to show 1 layer of the unit cell and indicate how the second layer is arranged.
(ii) Give the co-ordination numbers of the Na^+ and Cl^- ions.
(iii) Give the dimensions of the unit cell and the number of ion pairs it contains.
(iv) Calculate the volume share of 1 ion pair.
(v) Calculate the total volume of 1 mole of Na^+ and Cl^-.
(vi) Calculate Avogadro's number.

(b) Describe the crystal structure of caesium chloride giving the co-ordination numbers of Cs^+ and Cl^- ions. Explain why this structure differs from that of NaCl.

5 Solutions

1 *Miscible liquids*

(a) Ideal solutions

An ideal solution of 2 liquids A and B is one in which the forces between the molecules in the mixture of A and B are the same as those between the molecules in pure A and pure B.

Raoult's law applies to ideal solutions:

For a mixture of liquids, the partial vapour pressure of each component is proportional to its mole fraction in the mixture.

For the mixture of liquids A and B,

partial vapour pressure of A =

$$\frac{\text{number of moles } A}{(\text{number of moles } A + \text{number of moles } B)} \times \begin{array}{l}\text{vapour pressure} \\ \text{of pure } A \text{ at} \\ \text{temperature } T\end{array}$$

or

$$p_A = x_A p_A^\circ.$$

Similarly for component B

$$p_B = x_B p_B^\circ.$$

By Dalton's law,

total vapour pressure p above the mixture $= p_A + p_B$

$$\therefore p = x_B p_A^\circ + x_B p_B^\circ,$$

but

$$x_A + x_B = 1$$

$$\therefore p = x_A p_A^\circ + (1 - x_A) p_B^\circ.$$

The relationship between the vapour pressure and composition for the mixture at a particular temperature T is shown in the diagram below. At a particular composition W

the total vapour pressure $Z = WX + WY$.

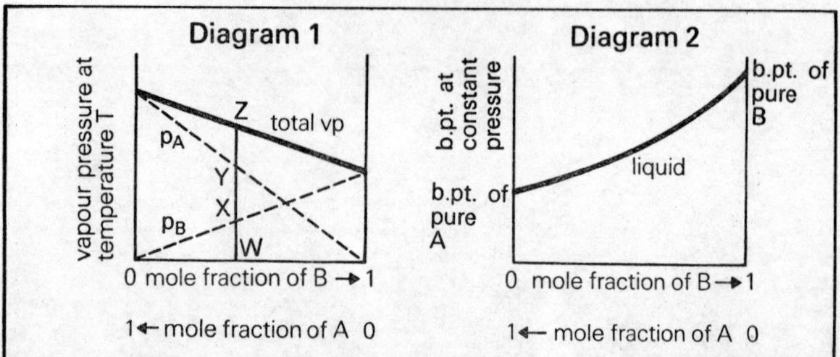

The boiling point composition graph (at constant pressure) has the opposite slope, as the higher the vapour pressure the lower the boiling point and vice versa. Graphs of these types are obtained for pairs of liquids which are chemically similar, e.g. heptane and hexane; iodoethane and bromoethane.

(b) **Non ideal solutions of liquids**

Most liquid solutions do not behave ideally as the forces between the molecules of *A* and *B* are not the same as those between the molecules in pure *A* or pure *B*. These liquid mixtures do not obey Raoult's law and so the relationship between the vapour pressure and molar composition is non-linear.

(i) **Negative deviation from Raoult's law**

The total vapour pressure above the mixture is less than expected because the attractive forces between molecules of *A* and *B* are greater than those between molecules in pure *A* or pure *B*.

On making the mixture there is often a contraction in volume and an evolution of heat, e.g. mixtures of water and mineral acids. The boiling point composition curve has a maximum.

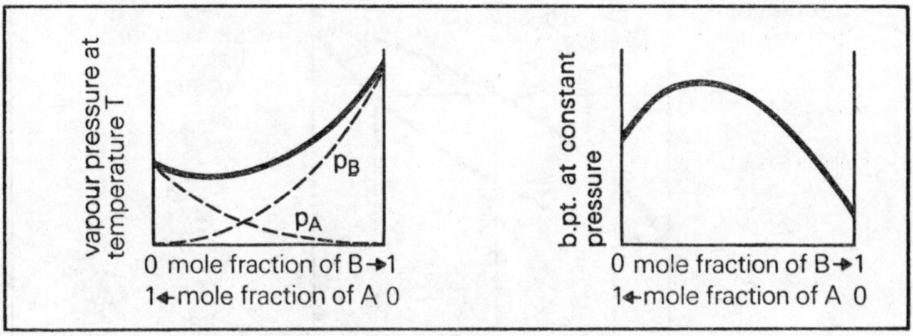

(ii) **Positive deviation from Raoult's law**

The total vapour pressure above the mixture is greater than expected because the attractive forces between the molecules of *A* and *B* are less than those between the molecules in pure *A* or pure *B*.

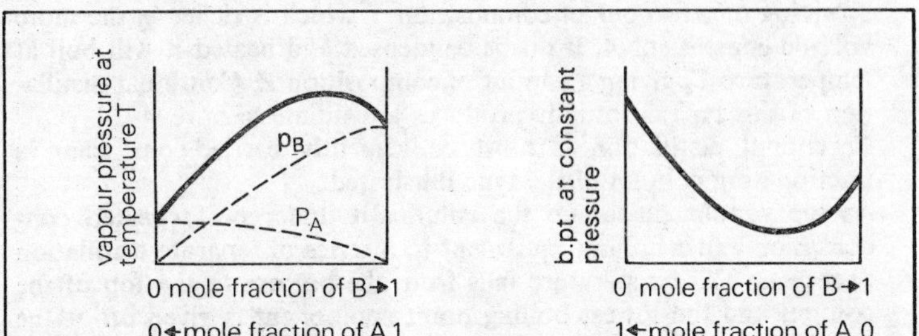

On making the mixture there is often an increase in volume and an absorption of heat, e.g. ethanol and water. The boiling point composition curve has a minimum.

In pure ethanol and pure water there are a large number of hydrogen bonds between the molecules. When the mixture is made these bonds are broken and replaced by a smaller number of hydrogen bonds between water and ethanol molecules.

Fractional distillation

(a) **Ideal solutions**

Ideal solutions of liquids are separated by fractional distillation. When the mixture of liquids is boiled, the vapour given off is richer in the more volatile constituent. The liquid line in the boiling point composition graph therefore has an associated vapour composition line lying above it. This is shown in the phase diagram below.

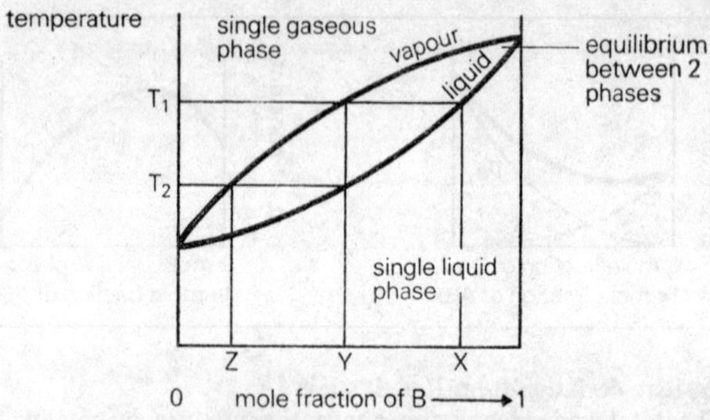

If a mixture of composition X is heated it will boil at temperature T_1 giving off a vapour of composition Y, which is richer in the more volatile constituent A. If this is condensed and heated it will boil at temperature T_2 giving a vapour of composition Z. Continual distillation of this type eventually produces a distillate of pure A.

Fractional distillation is most conveniently carried out using a fractionating column of the type illustrated.

As the vapour passes up the column it undergoes repeated condensation and reboiling equivalent to a series of separate distillation processes. The temperature falls from the bottom to the top of the column and the lowest boiling point component is given off at the top. The larger the number of trays the more efficient the separation.

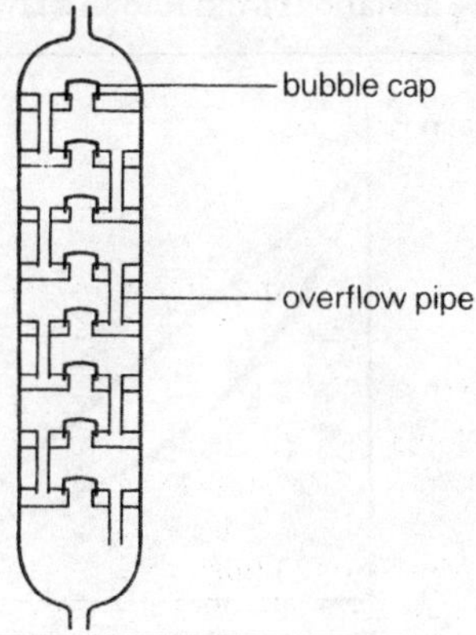

Mixtures of various composition can be drawn off at different points on the column.

(b) Non ideal solutions

(i) Negative deviation from Raoult's law, e.g. HNO_3, H_2O mixture.

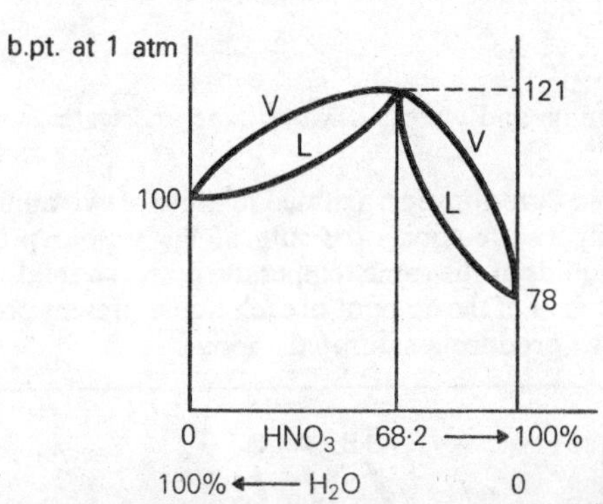

When a mixture containing less than 68·2% HNO_3 is fractionally distilled, the distillate is at first pure water. The mixture in the flask becomes more concentrated until it contains 68·2% HNO_3. It then distils unchanged giving a vapour containing 68·2% HNO_3. It is therefore called a constant boiling point mixture or azeotropic mixture. Mixtures containing more than 68·2% HNO_3 give pure HNO_3 on distillation until the residue has the composition of the azeotropic mixture. This then distils unchanged.

(ii) Positive deviations from Raoult's law, e.g. ethanol, water mixture.

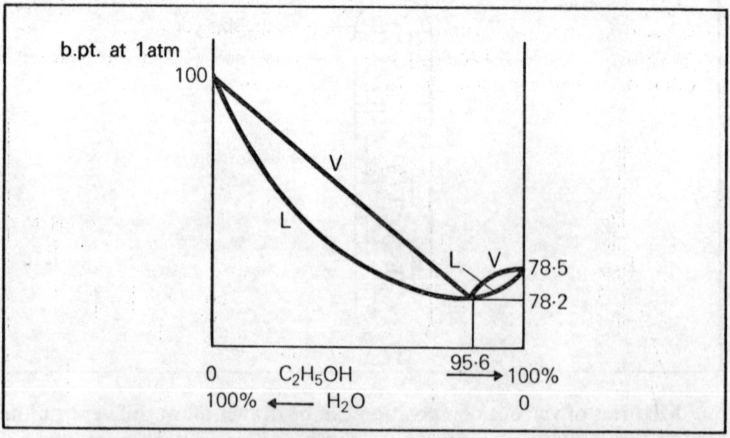

Mixtures containing more than 95·6% ethanol give pure ethanol and a minimum boiling point mixture of composition 95·6% ethanol on distillation. With mixtures containing less than 95·6% ethanol, water and the azeotropic mixture are obtained.

2 *Immiscible liquids*

For example phenylamine and water; ethoxyethane and water.

(a) **Steam distillation**

The total vapour pressure above an agitated mixture of two immiscible liquids at any temperature equals the sum of the vapour pressures of the individual liquids at the same temperature, i.e. the total vapour pressure is independent of the amount of each liquid present provided there is sufficient to produce a saturated vapour.

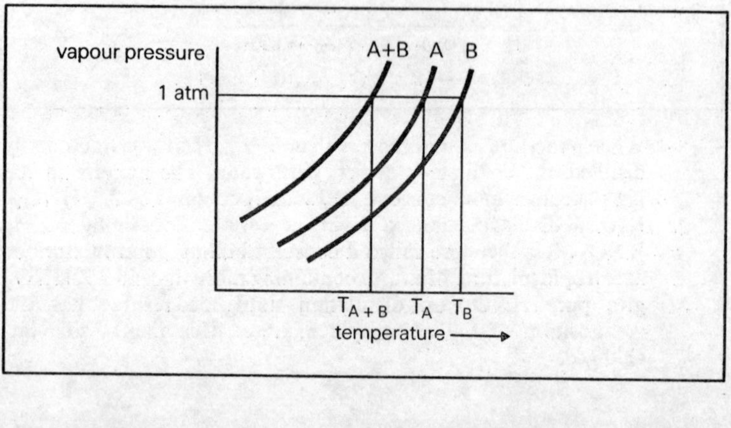

The boiling point of the mixture of 2 immiscible liquids *A* and *B* (T_{A+B}) is lower than that of pure $A(T_A)$ or pure $B(T_B)$. This property is used in steam distillation to obtain organic compounds from mixtures at temperatures below their boiling points. (Decomposition often occurs in the region of the boiling point).

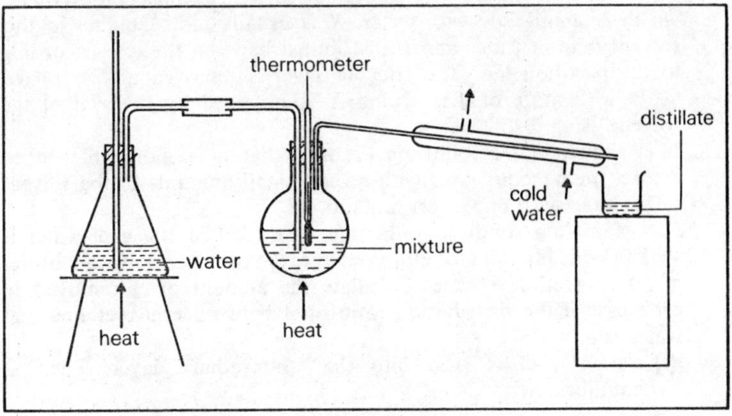

Steam distillation may be used to find the relative molecular mass of an organic compound *X*.
The number of moles of water and *X* present in the distillate is proportional to their vapour pressures at the temperature of boiling, i.e.

$$\frac{\text{vapour pressure of water}}{\text{vapour pressure } X} = \frac{\text{moles of water in the distillate}}{\text{moles of } X \text{ in the distillate}}$$

$$\therefore \frac{p_{H_2O}}{p_x} = \frac{\text{mass of water collected}}{18} \times \frac{M_r \text{ of } X}{\text{mass of } X \text{ collected}}.$$

If the temperature of distillation is recorded, then the vapour pressure of water at that temperature is found from tables.

$$\text{Then } p_x = \text{atmospheric pressure} - p_{H_2O}$$

The relative molecular mass of *X* is then calculated.

(b) **Distribution of solids, liquids and gases between immiscible liquids**

Solids and liquids

If a solid or liquid is added to a mixture of 2 immiscible liquids then it will distribute itself between them according to the distribution or partition law.
This states that **at equilibrium at constant temperature, the ratio of the concentrations of the solute in the 2 solvents is constant provided that the solute does not associate or dissociate in either,** i.e.

$$\frac{\text{concentration of } X \text{ in solvent } A}{\text{concentration of } X \text{ in solvent } B} = K$$

K is the distribution or partition constant.

The equilibrium established is dynamic.
This principle is used in the technique of solvent extraction for obtaining an organic solid or liquid X from its solution in water. The solution is shaken with an organic solvent, such as ethoxyethane, which is immiscible with water. X is usually more soluble in this solvent than in water and is distributed between the two according to the partition law. The aqueous layer is removed and extracted again with more of the solvent. X is obtained by removal of the solvent using distillation.
It is shown in the following example that it is more efficient to extract the aqueous layer with several small amounts of the solvent rather than with one larger amount.
For example a solution containing 10 g of A in 100 g of water is shaken with (a) 100 g of ethoxyethane, (b) two successive quantities of 50 g of ethoxyethane. Calculate the amount of A removed in each case if the distribution ratio of A between ethoxyethane and water is 5.

(i) Let x g of A pass into the ethoxyethane layer. Then at equilibrium

$$\frac{\frac{x}{100}}{\frac{(10 - x)}{100}} = 5$$

$$\therefore \quad x = 50 - 5x$$

$$6x = 50$$

$$x = 8{\cdot}33.$$

$\therefore$ The mass of A extracted $= 8{\cdot}33$ g (corr. to 3 sig. figs.).

(ii) **1st extraction** Let y g of A pass into the ethoxyethane layer. At equilibrium

$$\frac{\frac{y}{50}}{\frac{(10 - y)}{100}} = 5$$

$$\therefore \quad 2y = 50 - 5y$$

$$y = 7{\cdot}143.$$

$\therefore$ Mass of A left in the water layer $= 2{\cdot}857$ g.

2nd extraction Let z g of A pass into the ethoxyethane layer.

$$\frac{\frac{z}{50}}{\frac{(2{\cdot}857 - z)}{100}} = 5$$

$$\therefore \qquad 2z = 14{\cdot}285 - 5z$$

$$z = 2{\cdot}041.$$

$\therefore$ Total mass of A extracted = 7·143 + 2·041 = 9·18 g (corr. to to 3 sig. figs.).

Gases

The partition law can be applied to the distribution of a gas between air and water.

From the law it follows that

$$\frac{\text{concentration of gas in gaseous layer}}{\text{concentration of gas in water}} = K \text{ (a constant)}$$

or

$$\frac{\text{pressure of gas}}{\text{mass dissolved per cm}^3} = K^1 \text{ (a constant).}$$

This is an expression of **Henry's law** which states that **the mass of gas dissolved by a given volume of liquid at constant temperature is proportional to the applied pressure.** However, this law only applies if the gas does not associate or dissociate in the liquid. For mixtures of gases, Henry's law applies if each gas is treated separately and its partial pressure used.

Note: From Boyle's law it follows that the volume of gas dissolved, measured at the pressure used, is **independent** of the applied pressure. The solubility of a gas usually decreases with an increase in temperature.

3 *Solutions of solids in liquids*

On addition of a non-volatile solute to a solvent the vapour pressure of the solvent is lowered. If the solution is dilute, Raoult's law will be obeyed.

$$p = x_A p_A{}^\circ + x_B p_B{}^\circ$$

If B is the non-volatile solute then $p_B{}^\circ$ is zero. Then,

$$p = xp_0$$

where p is the vapour pressure of the solution; p_0 is the vapour pressure of the solvent; x is the mole fraction of the solvent.

$$\therefore \quad 1 - \frac{p}{p_0} = 1 - \text{mole fraction of solvent}$$

$$\frac{p_0 - p}{p_0} = \text{mole fraction of solute}$$

or

$$\frac{p_0 - p}{p_0} = \frac{n}{n + N}$$

where n is the number of moles of solute; N is the number of moles of solvent.

I.e. the relative lowering of the vapour pressure = mole fraction of the solute or, the relative lowering of the vapour pressure $\propto$ molar concentration of the solute.

Thus equimolecular quantities of any non-volatile solute dissolved in the same quantity of the same solvent will produce the same lowering of vapour pressure, providing that the solute does not associate or dissociate in solution. For example, 0·001 M glucose solution has the same vapour pressure as 0·001 M urea solution.

The lowering of the vapour pressure affects the freezing point, boiling point and osmotic pressure of the solution. The size of the effect depends on the number of solute particles present (moles of solute). These effects are known as the colligative properties of the solution.

Measurement of these properties of a solution enables the relative molecular mass of the solute to be found.

(a) M_r from vapour pressure measurements

By Raoult's law

$$\frac{p_0 - p}{p_0} = \frac{n}{N + n}$$

If the solution is dilute

$$\frac{p_0 - p}{p_0} \simeq \frac{n}{N}$$

or

$$\frac{p_0 - p}{p_0} = \frac{w_2 M_1}{M_2 w_1}$$

where w_1 and w_2 are the mass of the solvent and solute of relative molecular mass M_1 and M_2 respectively.

Thus M_2 can be calculated if the lowering of the vapour pressure of a solvent by a known mass of solute is found experimentally.

(b) *M* from boiling point measurements

The dissolution of a non-volatile solute in a solvent raises its boiling point.
Consider 2 solutions of the same non-volatile solute which does not associate or dissociate in the solvent. Solution ② is more concentrated than solution ①.

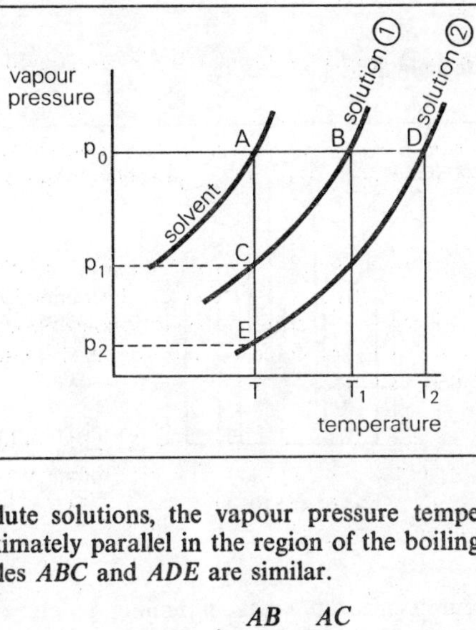

For dilute solutions, the vapour pressure temperature curves are approximately parallel in the region of the boiling point.
Triangles *ABC* and *ADE* are similar.

$$\therefore \frac{AB}{AD} = \frac{AC}{AE}$$

or

$$\frac{T_1 - T}{T_2 - T} = \frac{p_0 - p_1}{p_0 - p_2}$$

i.e. the elevation of the boiling point is proportional to the lowering of the vapour pressure.
From Raoult's law, equimolecular quantities of solutes dissolved in a given quantity of the same solvent produce the same lowering of vapour pressure and thus the same elevation of boiling point.
The **boiling point constant, molar elevation constant** or **ebullioscopic constant** K_b of a solvent is the elevation of boiling point produced when 1 mole of any solute is dissolved in 1000 g of that solvent. For water $K_b = 0{\cdot}52$ K kg^{-1} mol^{-1}, i.e. 1 mole of glucose dissolved in 1000 g of water raises the b.pt. by 0·52 K.

If w_2 g of solute of relative molecular mass M_2 is dissolved in w_1 g of solvent of boiling point constant K_b then by ratio the elevation of boiling point ΔT is given by

$$\Delta T = \frac{1000 K_b w_2}{w_1 M_2}.$$

∴ From experimental determination of ΔT, M_2 is found.
Note: Solutions must be dilute and the solute must not associate or dissociate.

Experimental technique

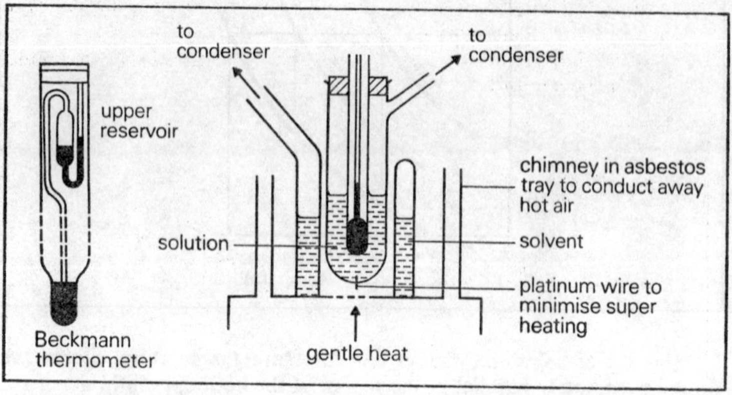

In making accurate measurements of boiling points and freezing points a Beckmann thermometer is used. It measures temperature changes but not absolute temperatures. The capillary is very fine so that a small temperature change causes a considerable movement of the mercury thread. The thermometer is first set so that for the temperature range involved the mercury will be on the scale. For measurements of high temperatures some of the mercury in the bulb is transferred to the upper reservoir and vice versa. In measuring boiling point elevations, apparatus similar to that shown may be used. The glass jacket containing hot solvent helps to ensure even heating. A known mass of pure solvent is placed in the inner container and its boiling point found. A weighed pellet of solute is introduced through the side arm. It is dissolved and the boiling point of the solution is found. Further pellets of known mass are added and the new boiling points recorded. A graph is plotted of the elevation of boiling point ΔT against the total mass of solute present w_2.

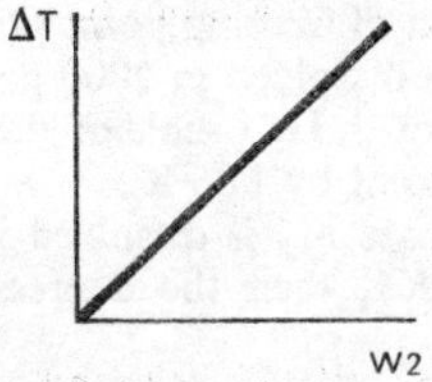

since $\Delta T = \dfrac{1000\,K_b\,w_2}{w_1 M_2}$

the gradient $= \dfrac{1000\,K_b}{w_1 M_2}$

M_2 can therefore be found

(c) *M_r* from freezing point measurements

The dissolution of a non-volatile solute in a solvent lowers its freezing point.

Consider 2 solutions of the same non-volatile solute which does not associate or dissociate in the solvent. Solution ② is more concentrated than solution ①. On freezing pure solvent only crystallizes out of the solution.

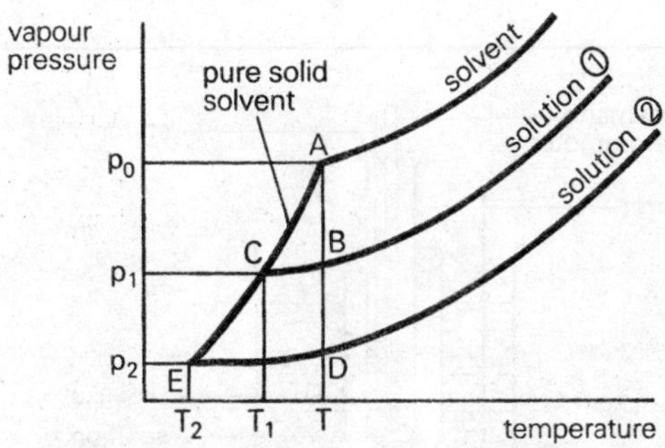

For dilute solutions the vapour pressure temperature curves are parallel in the region of the freezing point.

Triangles *ABC* and *ADE* are similar.

$$\frac{AB}{AD} = \frac{BC}{DE}$$

$$\frac{p_0 - p_1}{p_0 - p_2} = \frac{T - T_1}{T - T_2}$$

i.e. the depression of the freezing point is proportional to the lowering of the vapour pressure.

The **freezing point constant, molar depression constant or cryoscopic constant** K_f of a solvent is the depression of freezing point produced when 1 mole of a non-volatile solute is dissolved in 1000 g of that solvent. For water $K_f = 1{\cdot}86$ K kg^{-1} mol^{-1}, i.e. 1 mole of glucose in 1000 g of water depresses the freezing point by 1·86 K.
If w_2 g of solute of relative molecular mass M_2 is dissolved in w_1 g of solvent of freezing point constant K_f, then the depression of freezing point ΔT is given by

$$\Delta T = \frac{1000 K_f w_2}{w_1 M_2}$$

$\therefore$ From experimental determinations of ΔT, M_2 is found. The solutions must be dilute and the solute must not associate or dissociate.

Experimental technique

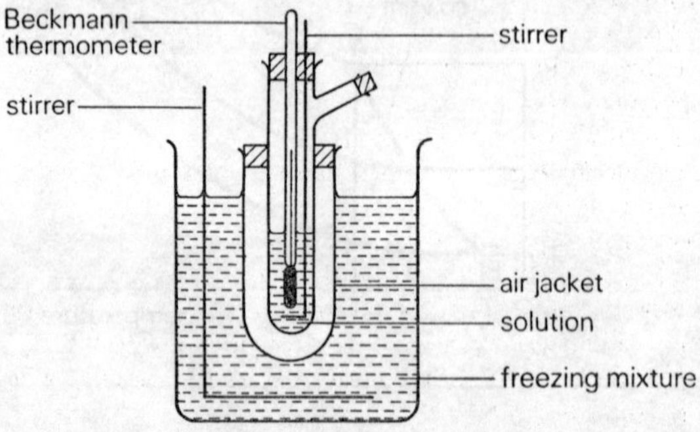

Apparatus similar to that shown may be used. A known mass of solvent is placed in the inner vessel and cooled. A cooling curve is plotted and the freezing point found.
A weighed pellet of solute is dissolved in the melted solvent and the solution is cooled. A cooling curve is again plotted and the new freezing point found. Pure solvent crystallizes out on cooling so the solution becomes more concentrated and the freezing point falls. Further pellets of known mass are added and the new freezing points found.

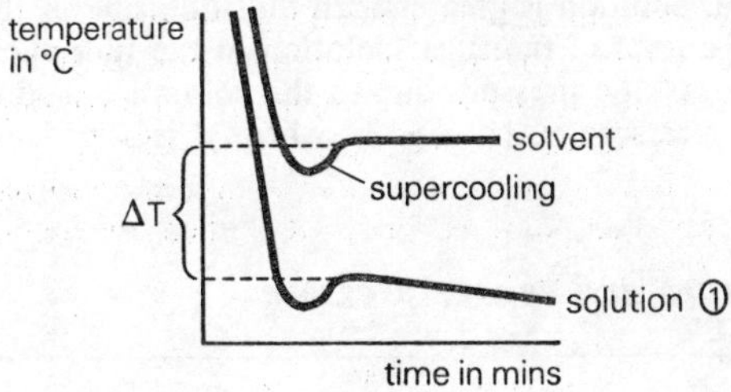

Cooling curves

A graph is plotted of the depression of freezing point, ΔT, against the total mass of solute added, w_2. Since

$$\Delta T = \frac{1000 K_f w_2}{w_1 M_2}$$

$$\text{the gradient} = \frac{1000 K_f}{w_1 M_2}$$

M_2 can therefore be found.

(d) **M_r from osmotic pressure measurements**

Osmosis is the diffusion of solvent molecules through a semi-permeable membrane from a solvent or dilute solution into a more concentrated solution. A **semi-permeable membrane** is a material which allows solvent but not solute molecules to pass through. Many animal membranes, such as bladders, are approximately semi-permeable but are physically very weak. The most suitable material for use in osmotic pressure measurements is copper(II) hexacyano-ferrate(II) deposited on the walls of a porous pot.

The **osmotic pressure** of a solution is the pressure which must be exerted on the solution to prevent osmosis from taking place. When osmosis has stopped, diffusion of solvent molecules still occurs but the rate is equal in both directions.

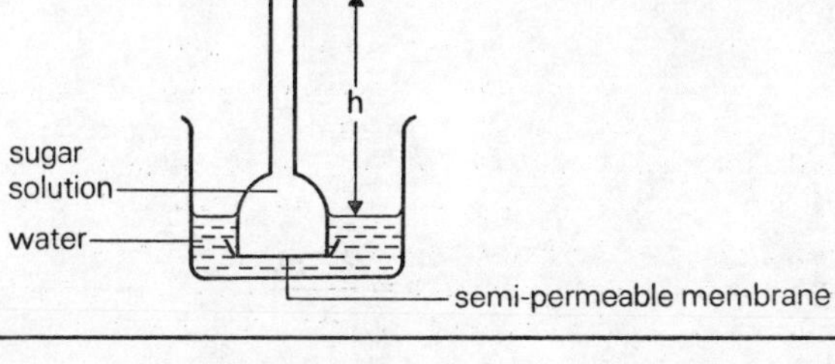

At the beginning of the experiment, the rate of diffusion of water into the sugar solution is greater than that diffusing in the opposite direction. The level of the sugar solution in the tube therefore rises until the hydrostatic pressure due to the column of solution equals the osmotic pressure π, i.e. $\pi = hd$ where d is the density of the solution.

Osmotic pressure and vapour pressure

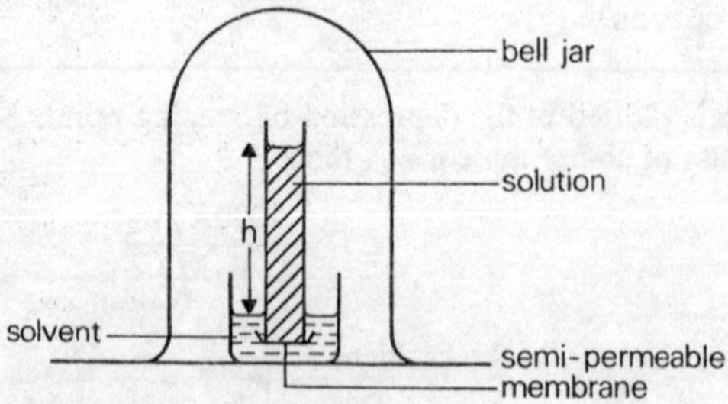

The bell jar is first evacuated so that at equilibrium it contains solvent vapour only. At equilibrium

$$\pi = hd_1$$

where d_1 = density of the solution. For dilute solutions

$$d_1 = d$$

where d is the density of the solvent.

$$\therefore \pi = hd$$

If p_0 is the vapour pressure of the pure solvent, p the vapour pressure of the solution at the top of the column and ρ the average density of the vapour, then

$$p_0 - p = h\rho$$

or

$$p_0 - p = \frac{\pi\rho}{d}$$

but

$$p = \frac{M_r p_0}{RT}$$

where M_r is the relative molecular mass of the solvent.

$$\therefore p_0 - p = \frac{\pi M_r p_0}{dRT}$$

$$\frac{p_0 - p}{p_0} \propto \pi$$

Since the osmotic pressure is proportional to the relative lowering of the vapour pressure, it is a colligative property of the solution.

Laws of osmosis

The following relationships are found to apply to dilute solutions of solutes which do not associate or dissociate in solution.

1 Concentration effect

At constant temperature, the osmotic pressure of a dilute aqueous solution of a solute is proportional to its concentration, i.e. $\pi \propto C$.
For a given mass of solute, the concentration is inversely proportional to the volume, V, of the solution, i.e. $\pi \propto 1/V$.
This law is analogous to Boyle's law for ideal gases.

2 Temperature effect

For a given solution the osmotic pressure is proportional to the absolute temperature, T, i.e. $\pi \propto T$.
This law is analogous to Charles' law for ideal gases.
On combining these laws the following general equation is obtained.

$$\pi V = kT$$

(k is a constant for a given mass of solute).
If 1 mole of solute is used

$$\pi V = RT$$

where R is the universal gas constant.
For n moles of solute

$$\pi V = nRT$$

(see page 25 for units and relevant values of R).
The following relationship also applies

$$\frac{\pi_1}{C_1 T_1} = \frac{\pi_2}{C_2 T_2}$$

where π_1, C_1 and T_1 are the initial osmotic pressure, concentration and temperature of the solution; π_2, C_2 and T_2 are the final osmotic pressure, concentration and temperature of the solution.
Note: A solution containing 1 mole of solute dissolved in 22·4 l of solution at 0°C exerts an osmotic pressure of 1 atmosphere (101325 N m^{-2}). Equimolecular solutions of different solutes in the same solvent have the same osmotic pressure at the same temperature. They are said to be **isotonic** (no osmosis will occur between them).

Experimental technique

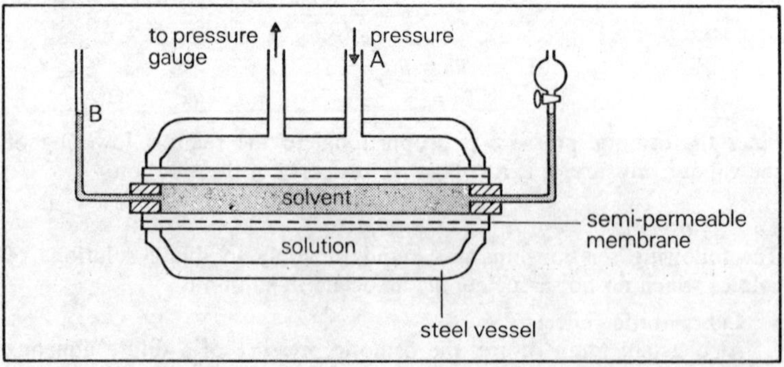

A pressure is applied at *A* to just balance the osmotic pressure of the solution. The level of solvent in *B* is then constant. In practice exact balancing is impossible so the average is taken of the pressures which cause inflow to and allow outflow from the solvent. Osmotic pressure measurements are used to find the relative molecular mass of large molecules such as polymers. These produce too small a freezing point depression or elevation of boiling point for these measurements to be sufficiently accurate.
For example:

(a) A 1·5% solution of a polymer in dioxan had an osmotic pressure of 0·0026 atms at 20°C. What is the relative molecular mass of the polymer if $R = 0{\cdot}0821$ l atm K^{-1} mol^{-1}?

Using $$\pi V = \frac{m}{M_r} RT$$

$$M_r = \frac{mRT}{\pi V}$$

$$= \frac{1{\cdot}5 \times 0{\cdot}0821 \times 293}{0{\cdot}0026 \times 0{\cdot}1}$$

$$= 1{\cdot}39 \times 10^5$$

∴ The relative molecular mass of the polymer is $1{\cdot}39 \times 10^5$ (corr. to 3 sig. figs.).

(b) Calculate the osmotic pressure of an aqueous solution containing 12 g of urea (M_r 60) in 500 cm^3 at 12°C.

$$\text{Using } \frac{\pi_1}{C_1 T_1} = \frac{\pi_2}{C_2 T_2}$$

where $\pi_1 = ?$ $\quad C_1 = \frac{12}{500}$ g cm^{-3} $\quad T_1 = 285$ K

$\pi_2 = 1$ atm $\quad C_2 = \frac{60}{22400}$ g cm^{-3} $\quad T_2 = 273$ K

$$\frac{\pi_1}{\frac{12}{500} \times 285} = \frac{1}{\frac{60}{22400} \times 273}$$

$$\therefore \pi_1 = 9{\cdot}35$$

∴ The pressure of the solution is 9·35 atm (corr. to 3 sig. figs.).

Dissociation and association

(a) **Dissociation** If a solute dissociates in solution then more particles than expected will be present and the colligative effect will be larger. The value of the relative molecular mass calculated from this measurement will therefore be smaller than expected. For example a solution containing 1 mole of sodium chloride in 1000 g of water has a freezing point of $-3{\cdot}73$°C. Since K_f for water is 1·86 K kg^{-1} mol^{-1} complete dissociation must have occurred.

$$NaCl \longrightarrow Na^+ + Cl^-$$

A solution containing 1 mole of calcium chloride in 1000 g of water has a freezing point of $-5{\cdot}58$°C. Complete dissociation occurred.

$$CaCl_2 \longrightarrow Ca^{2+} + 2Cl^-$$

Many solutes do not dissociate completely. The degree of dissociation, α, can be calculated from the observed and expected colligative effect. Consider 1 mole of a solute A dissociating into n ions in solution.

	A	$\longrightarrow$	n ions
No. of moles before dissociation	1		0
No. of moles after dissociation	$1 - \alpha$		$n\alpha$

The observed colligative effect is proportional to the number of moles present.

$$\frac{\text{Expected effect if no dissociation occurs}}{\text{Observed effect}} = \frac{1}{1 + \alpha(n-1)}$$

But the colligative effect is inversely proportional to the M_r of the solute.

$$\therefore \frac{\text{Expected } M_r}{\text{Apparent } M_r} = \frac{1 + \alpha(n-1)}{1}$$

This ratio is sometimes known as van't Hoff's factor i.

(b) **Association** If association occurs then the observed colligative effect is smaller and the relative molecular mass calculated from this value is larger than expected. For example 1 mole of ethanoic acid in 1000 g of benzene depresses the freezing point by 2·56 K (K_f for benzene is 5·12 K kg^{-1} mol^{-1}).

$\therefore$ number of moles present = 2·56/5·12 = 0·5.

Association has occurred producing dimers.

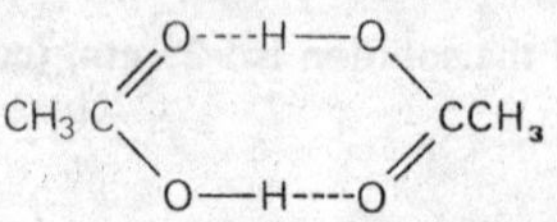

Calculation
A solution of chloroethanoic acid containing 5·57 g in 500 cm^3 of aqueous solution is isotonic with a cane sugar solution ($C_6H_{12}O_{11}$) containing 4·05 g in 100 cm^3 of aqueous solution. Calculate the percentage dissociation of the acid.

Isotonic solutions have the same osmotic pressure and therefore contain the same number of particles per unit volume of solution.
Concentration of sugar solution = 4·05/(342 × 100) mol cm^{-3}.
Let the apparent M_r of the acid be x.
Then concentration of the acid solution = 5·57/(x × 500) mol cm^{-3}.

But $$\frac{5{\cdot}57}{x \times 500} = \frac{4{\cdot}05}{342 \times 100}$$

$$\therefore \qquad x = \frac{5{\cdot}57 \times 342}{5 \times 4{\cdot}05} = 94{\cdot}07.$$

In aqueous solution $CH_2ClC(=O)OH \rightleftharpoons CH_2ClC(=O)O^- + H^+$

	$CH_2ClCOOH$	CH_2ClCOO^-	H^+
No. of moles before dissociation	1	0	0
No. of moles after dissociation	$(1-\alpha)$	α	α

$$\frac{\text{Calculated } M_r}{\text{Observed } M_r} = \frac{1+\alpha}{1} = \frac{94{\cdot}5}{94{\cdot}07}$$

$$\therefore \alpha = \frac{94{\cdot}5}{94{\cdot}07} - 1$$

$$= \frac{0{\cdot}43}{94{\cdot}07}$$

$$= 0{\cdot}00457$$

∴ The percentage dissociation of the acid = 0·457% (corr. to 3 sig. figs.).

Practice questions

In questions 1 to 10 give the letter of the correct answer.

1 Which of the following in 0·1 *M* aqueous solution will have the lowest vapour pressure at 298 K?
A. KOH B. HNO_3 C. C_2H_5OH D. $CaCl_2$ E. CH_3COOH

2 Which of the following statements about an azeotropic mixture of 2 liquids must be true?
A. At its b.pt. the composition of the azeotropic mixture is the same as that of its vapour.
B. The v.p. of the azeotropic mixture is less than that of either of its components.
C. The b.pt. of the azeotropic mixture is greater than that of either of its components.
D. The composition of the azeotropic mixture is independent of the external pressure.
E. The composition of the mixture cannot be altered by any process involving distillation.

3 The distribution (partition) coefficient for a compound *X* between benzene and water is 3. 100 cm^3 of an aqueous solution containing 10 g of *X* is shaken with 100 cm^3 of benzene. The mass of *X* in g left in the aqueous layer is
A. $7\frac{1}{2}$ B. $6\frac{2}{3}$ C. $3\frac{1}{3}$ D. 3 E. $2\frac{1}{2}$

4 2 immiscible liquids *X* and *Y* are placed in a stoppered bottle. If the mole fraction of *X* is *x* and the saturation vapour pressures of *X* and *Y* are p_x and p_y respectively, the total vapour pressure in the bottle at equilibrium is
A. $p_x + p_y$ B. $(p_x + p_y)/2$ C. $xp_x + (1 - x)p_y$
D. $(1 - x)p_x + p_y$ E. $2(p_x + p_y)$

5 0·1 *M* aqueous glucose solution freezes at −0·185°C. The freezing point of 0·1 *M* aqueous ethanoic acid will be
A. −0·093 B. −0·185 C. −0·180 D. −0·190 E. −0·370

6 Which of the following in aqueous solution has a b.pt. most equal to that of an equimolar solution of $K_4[Fe(CN)_6]$?
A. NH_4NO_3 B. CCl_4 C. Na_2SO_4 D. $Al_2(SO_4)_3$ E. $AlCl_3$

7 Which of the following when dissolved in benzene at a concentration of 0·5 g l^{-1} would give the solution with greatest osmotic pressure?
A. $C_6H_5COCH_3$ B. poly(chloroethene) C. poly(phenylethene)
D. $C_6H_{12}O_6$ E. $CO(NH_2)_2$

8 Which of the following statements about the steam distillation of a substance *X* is correct?
A. *X* must be miscible with water.
B. *X* distils over at its normal b.pt.
C. For efficiency M_r of *X* must be small.
D. For efficiency, the v.p. of *X* must be small.
E. At the b.pt., the total v.p. of the mixture equals the external pressure.

9 Which of the following does not influence the v.p. of a solution of a non-volatile solute in a solvent?
A. The b.pt. of the solute.
B. The temperature of the solution.
C. The mole fraction of the solute.
D. The mole fraction of the solvent.
E. The degree of dissociation of the solute.

10 The molar solubility of oxygen in water is about twice that of nitrogen at the same temperature and pressure. Assuming air contains 20% oxygen by volume, the volume percentage of oxygen in the air dissolved in water is
A. 20 B. 40 C. 33 D. 67 E. 60

11 (a) State the distribution law for the partition of a solute between two immiscible solvents. Give, with reasons, one case in which the law is not obeyed.
(b) 50 cm^3 of a 0·3 *M* dibasic acid were shaken with 50 cm^3 of ethoxyethane and allowed to reach equilibrium. 10 cm^3 of the aqueous layer required 25 cm^3 of 0·2 *M* NaOH for complete neutralization. Calculate the distribution coefficient for the acid assuming the solvents are immiscible.
(c) A liquid *Y* and water are immiscible. A mixture of *Y* and water boils at 95°C and 760 mm of mercury. At 95°C, water has a v.p. of 640 mm of mercury. The distillate contains $Y : H_2O$ in a ratio 1·65 : 1 by mass. Find M_r of *Y*.

12 Draw curves to show variation of the v.p. of a mixture of two miscible, volatile solvents with the composition expressed in mole fractions. (Consider the ideal and 2 non-ideal cases giving reasons for the curves.) In each case draw the b.pt. composition curve. Show how the curves may be used to explain fractional distillation and, in one non-ideal case only, the existence of an azeotropic mixture.

13

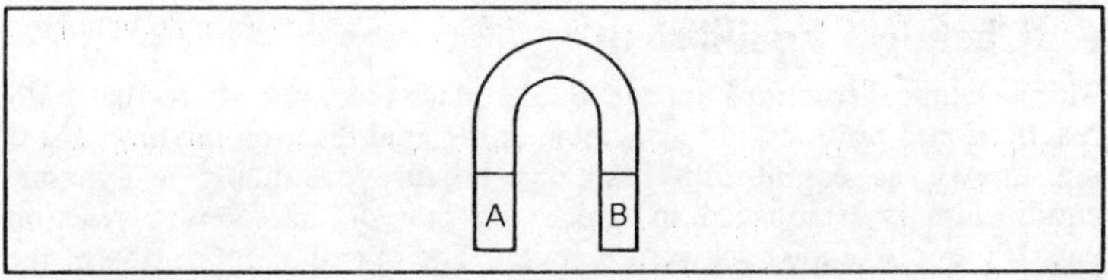

Each arm of the sealed *U*-tube contains 50 cm^3 of a KCl solution but the concentration in *A* is twice that in *B*. Describe quantitatively what you would observe after equilibrium has been established. Give reasons for your answer.

6 Chemical Equilibrium

Many chemical reactions appear to take place incompletely so that both reactions and products are detectable in the final reaction mixture. These are known as equilibrium reactions because eventually a dynamic equilibrium is established in which the rate of the forward reaction equals the rate of the backward reaction. The overall composition of the reaction mixture then remains constant.

The equilibrium law

For any system in equilibrium it is found that a numerical relationship exists between the concentration of the reactants and products present. For the generalized reaction

$$aA + bB \rightleftharpoons cC + dD$$

where a and b are the number of moles of reactants A and B and c and d are the number of moles of reactants C and D.

The equilibrium law states that at constant temperature

$$K_c = \frac{[C]^c[D]^d}{[A]^a[B]^b}$$

where [ ] denotes the molar concentration at equilibrium in mol l^{-1} and K_c is the equilibrium constant in terms of concentration.

K_c is constant for any particular equilibrium at a specified temperature. If the total number of moles of products in the balanced equation equals the total number of moles of reactants then K_c is dimensionless.
For gaseous reactions it is often more convenient to express the concentration in terms of partial pressures.
For the generalized gaseous reaction

$$aA + bB \rightleftharpoons cC + dD$$

$$K_p = \frac{p_C{}^c p_D{}^d}{p_A{}^a p_B{}^b}$$

where p_A is the partial pressure of A in the equilibrium mixture etc. and K_p is the equilibrium constant in terms of partial pressures.

For gaseous reactions in which the number of moles of products equals the number of moles of reactants, $K_p = K_c$.

Homogeneous equilibria

Homogeneous equilibria are equilibrium reactions in which the products and reactants are in the same physical state.

1 Gaseous equilibria

(a) The hydrogen, iodine, hydrogen iodide system

(i) **Calculation of K_c** Consider a mixture of a moles of iodine and b moles of hydrogen in a volume of V litres at a particular temperature. Let x be the number of moles of iodine converted into hydrogen iodide under the equilibrium conditions.

	I_2 +	H_2 ⇌	2HI
Moles initially	a	b	0
Moles at equilibrium	$a - x$	$b - x$	$2x$
Equilibrium concentration mol l^{-1}	$\frac{(a-x)}{V}$	$\frac{(b-x)}{V}$	$\frac{2x}{V}$

$$K_c = \frac{[HI]^2}{[I_2][H_2]}$$

$$= \frac{\frac{4x^2}{V^2}}{\frac{(a-x)}{V}\frac{(b-x)}{V}}$$

$$= \frac{4x^2}{(a-x)(b-x)}$$

(ii) **Calculation of K_p** Consider the same mixture as above and let the total pressure be p.

	I_2 +	H_2 ⇌	2HI
Moles initially	a	b	0
Moles at equilibrium	$a - x$	$b - x$	$2x$

At equilibrium:

$$\text{Total number of moles} = a - x + b - x + 2x = a + b$$

$$\therefore \; p_{I_2} = \frac{(a-x)}{(a+b)}p \quad p_{H_2} = \frac{(b-x)}{(a+b)}p \quad p_{HI} = \frac{2x}{(a+b)}p$$

$$K_p = \frac{p^2_{HI}}{p_{I_2}p_{H_2}} = \frac{4x^2p^2(a+b)^2}{(a+b)^2(a-x)(b-x)p^2}$$

$$= \frac{4x^2}{(a-x)(b-x)}$$

Thus for this particular reaction $K_c = K_p$.

(b) Ammonia equilibrium

(i) **Calculation of K_c** Consider a mixture of 1 mole of nitrogen and 3 moles of hydrogen in a volume of V litres at a particular temperature. Let α be the fraction of a mole of nitrogen, which is converted into ammonia under the equilibrium conditions.

	N_2 +	$3H_2$ $\rightleftharpoons$	$2NH_3$
Moles initially	1	3	0
Moles at equilibrium	$1-\alpha$	$3(1-\alpha)$	2α
Equilibrium concentration mol l^{-1}	$\dfrac{(1-\alpha)}{V}$	$\dfrac{3(1-\alpha)}{V}$	$\dfrac{2\alpha}{V}$

$$K_c = \frac{[NH_3]^2}{[N_2][H_2]^3}$$

$$= \frac{4\alpha^2 V^4}{V^2 27(1-\alpha)^4}$$

$$K_c = \frac{4\alpha^2 V^2}{27(1-\alpha)^4}$$

K_c is constant at constant temperature and has the units of l^2.

Under normal conditions, α is small $\therefore$ $(1-\alpha)^4 \simeq 1$.

$$\therefore K_c \simeq \frac{4\alpha^2 V^2}{27} \qquad \therefore \alpha^2 \propto \frac{1}{V^2}$$

but by Boyle's law $p \propto 1/V$

$$\therefore \alpha^2 \propto p^2$$

or

$$\alpha \propto p$$

$\therefore$ The yield of ammonia is proportional to the applied pressure. This is experimentally confirmed.

(ii) **Calculation of K_p** Consider the same mixture as above and let the total pressure be p.

	N_2 +	$3H_2$ $\rightleftharpoons$	$2NH_3$
Moles initially	1	3	0
Moles at equilibrium	$1-\alpha$	$3(1-\alpha)$	2α

At equilibrium:

$$\text{Total number of moles} = 1-\alpha+3(1-\alpha)+2\alpha = 4-2\alpha$$

$$p_{N_2} = \frac{(1-\alpha)}{(4-2\alpha)}p \quad p_{H_2} = \frac{3(1-\alpha)}{(4-2\alpha)}p \quad p_{NH_3} = \frac{2\alpha}{(4-2\alpha)}p$$

$$K_p = \frac{p_{NH_3}{}^2}{p_{N_2}\,p_{H_2}} = \frac{4\alpha^2 p^2 (4-2\alpha)(4-2\alpha)^3}{(4-2\alpha)^2(1-\alpha)p27(1-\alpha)^3 p^3}$$

$$\therefore K_p = \frac{4\alpha^2(4-2\alpha)^2}{27(1-\alpha)^4 p^2}$$

K_p is constant at constant temperature and has the units of atm^{-2}. $K_p \neq K_c$ as the number of moles of products $\neq$ the number of moles of reactants.

If α is small, $(1 - \alpha) \simeq 1$ and $(4 - 2\alpha) \simeq 4$.

Then $$\alpha^2 \propto p^2$$

or $$\alpha \propto p$$

The yield of ammonia is shown to be proportional to the applied pressure.

2 *Liquid equilibria* For example esterification
Consider a mixture of a moles of ethanoic acid and b moles of ethanol in a volume of V litres. Let x moles of ethyl ethanoate be present at equilibrium at temperature T.

	CH_3COOH +	$C_2H_5OH \rightleftharpoons$	$CH_3COOC_2H_5$ +	H_2O
Moles initially	a	b	0	0
Moles at equilibrium	$(a - x)$	$(b - x)$	x	x
Equilibrium concentration $mol\ l^{-1}$	$\frac{a - x}{V}$	$\frac{b - x}{V}$	$\frac{x}{V}$	$\frac{x}{V}$

$$K_c = \frac{[CH_3COOC_2H_5][H_2O]}{[CH_3COOH][C_2H_5OH]} = \frac{x^2}{(a - x)(b - x)}$$

For example calculate the number of moles of ethyl ethanoate formed when 2 moles of ethanol and 1 mole of ethanoic acid are allowed to reach equilibrium if $K_c = 4$. Let the volume of the mixture be V and the number of moles of ethyl ethanoate formed be x.

	C_2H_5OH +	$CH_3COOH \rightleftharpoons$	$CH_3COOC_2H_5$ +	H_2O
Moles initially	2	1	0	0
Equilibrium concentration $mol\ l^{-1}$	$\frac{(2 - x)}{V}$	$\frac{(1 - x)}{V}$	$\frac{x}{V}$	$\frac{x}{V}$

$$K_c = \frac{[CH_3COOC_2H_5][H_2O]}{[C_2H_5OH][CH_3COOH]} = 4$$

$$\therefore \quad 4 = \frac{x^2}{(2 - x)(1 - x)}$$

$$\therefore 4(2 - 3x + x^2) = x^2$$

$$\therefore 3x^2 - 12x + 8 = 0$$

$\therefore K_c = 3{\cdot}155$ or $0{\cdot}845$

The former value is not possible. $\therefore$ The number of moles of ethyl ethanoate formed $= 0{\cdot}845$.

Heterogeneous equilibria

Heterogeneous equilibria are equilibrium reactions in which the products and reactants are in different phases. The concentrations of any solids or liquids present are taken as constant.

1 Thermal dissociation of carbonates
For example:

$$CaCO_3(s) \rightleftharpoons CaO(s) + CO_2(g)$$

The following relationship may be written.

$$K = \frac{[CaO(s)][CO_2(g)]}{[CaCO_3(s)]}$$

or

$$[CO_2(g)] = K\frac{[CaCO_3(s)]}{[CaO(s)]}$$

The concentration of calcium carbonate and calcium oxide are considered to be constant. Thus the right-hand side of the equation is a constant and by convention this is the value of K_c for the equilibrium, i.e.

$$K_c = [CO_2(g)]$$

also

$$K_p = p_{CO_2}$$

K_p is known as the dissociation pressure of calcium carbonate and is constant at constant temperature, i.e. at a particular temperature there is a certain pressure of carbon dioxide in equilibrium with the 2 solid phases and independent of the amounts of solids present. It increases with temperature as shown.

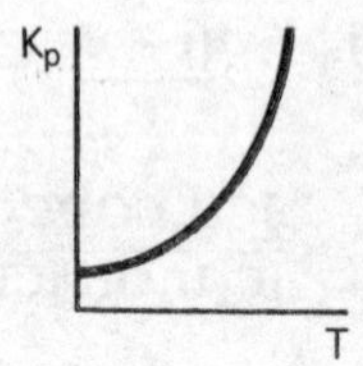

2 Vaporization of liquids
For example:

$$Br_2(l) \rightleftharpoons Br_2(g)$$

The concentration of liquid bromine is a constant and ∴ at any particular temperature $K_p = p_{Br_2}$.

3 Reaction between iron and steam
Above 100°C in a closed system the following equilibrium is set up

$$3Fe(s) + 4H_2O(g) \rightleftharpoons Fe_3O_4(s) + 4H_2(g)$$

$$\therefore K_p = \frac{p_{H_2}{}^4}{p_{H_2O}{}^4}$$

e.g. at a fixed temperature, the equilibrium constants K_p for the following equilibria are as given.

$$① \; Fe(s) + CO_2(g) \rightleftharpoons FeO(s) + CO(g)$$

$$K_p = 1{\cdot}47$$

$$② \; Fe(s) + H_2O(g) \rightleftharpoons FeO(s) + H_2(g)$$

$$K_p = 2{\cdot}38$$

To calculate K_p for the equilibrium

$$CO_2(g) + H_2(g) \rightleftharpoons CO(g) + H_2O(g) \; ③$$

For reaction ① $\quad K_p = \dfrac{p_{CO}}{p_{CO_2}} = 1{\cdot}47$

For reaction ② $\quad K_p = \dfrac{p_{H_2}}{p_{H_2O}} = 2{\cdot}38$

For reaction ③ $\quad K_p = \dfrac{p_{CO}p_{H_2O}}{p_{CO_2}p_{H_2}} = \dfrac{1{\cdot}47}{2{\cdot}38} = 0{\cdot}62$

4 Heterogeneous equilibria involving ions
The equilibrium law can be applied only to dilute solutions of ions. For more concentrated solutions, the forces between the ions become significant.

(a) Reactions between metals and metal ions
For example:

$$Cu(s) + 2Ag^+(aq) \rightleftharpoons Cu^{2+}(aq) + 2Ag(s)$$

$$K_c = \frac{[Cu^{2+}(aq)]}{[Ag^+(aq)]^2}$$

(b) Dissolved ions in equilibrium with solids. Solubility products
For example:

$$AgCl(s) + aq \rightleftharpoons Ag^+(aq) + Cl^-(aq)$$

The following relationship may be written

$$K = \frac{[Ag^+(aq)][Cl^-(aq)]}{[AgCl(s)]}$$

But [AgCl(s)] is constant,

$$\therefore K[\text{AgCl(s)}] = [\text{Ag}^+(\text{aq})][\text{Cl}^-(\text{aq})] = Ksp$$

Ksp is known as the solubility product and is constant at constant temperature. It is defined for sparingly soluble salts only, as their saturated solutions are sufficiently dilute for the equilibrium law to be applied.

For $Ca_3(PO_4)_2$

$$Ksp = [\text{Ca}^{2+}(\text{aq})]^3[\text{PO}_4{}^{3-}(\text{aq})]^2$$

In general, for the sparingly soluble salt $C_m A_n$,

$$Ksp = [\text{C}^{n+}(\text{aq})]^m[\text{A}^{m-}(\text{aq})]^n$$

If a solution containing C^{n+}(aq) ions is added to one containing A^{m-}(aq) ions then precipitation of the salt occurs when the solubility product is exceeded.

Calculation

Calculate the solubility in g l^{-1} at 298 K of silver chloride in (i) water, (ii) 0·1 *M* HCl. *Ksp* of AgCl = $1{\cdot}8 \times 10^{-10}$ mol^2 l^{-2}

(i) In a saturated aqueous solution

$$[\text{Ag}^+(\text{aq})] = [\text{Cl}^-(\text{aq})]$$

$$\begin{aligned} Ksp &= [\text{Ag}^+(\text{aq})][\text{Cl}^-(\text{aq})] \\ &= [\text{Ag}^+(\text{aq})]^2 \\ &= 1{\cdot}8 \times 10^{-10} \end{aligned}$$

$$\begin{aligned} \therefore [\text{Ag}^+(\text{aq})] &= \sqrt{1{\cdot}8 \times 10^{-10}} \\ &= 1{\cdot}34 \times 10^{-5} \end{aligned}$$

$$M_r \text{ AgCl} = 143{\cdot}5$$

$$\begin{aligned} \therefore \text{Solubility of AgCl} &= 143{\cdot}5 \times 1{\cdot}34 \times 10^{-5} \\ &= 1{\cdot}9 \times 10^{-3} \text{ g l}^{-1}. \end{aligned}$$

The solubility of silver chloride in water at 298 K = $1{\cdot}9 \times 10^{-3}$ g l^{-1}.

(ii) The contribution to the chloride ion concentration from the silver chloride is ignored as that from the hydrochloric acid is very large, i.e.

$$[\text{Cl}^-(\text{aq})] = 0{\cdot}1 \text{ g ions l}^{-1}$$

$$\begin{aligned} \therefore [\text{Ag}^+(\text{aq})] &= \frac{1{\cdot}8 \times 10^{-10}}{0{\cdot}1} \\ &= 1{\cdot}8 \times 10^{-9} \end{aligned}$$

$$\begin{aligned} \therefore \text{The solubility of AgCl} &= 143{\cdot}5 \times 1{\cdot}8 \times 10^{-9} \\ &= 2{\cdot}6 \times 10^{-7} \text{ g l}^{-1} \end{aligned}$$

∴ The solubility of silver chloride in 0·1 M HCl at 298 K is $2{\cdot}6 \times 10^{-7}$ g l^{-1}.

(c) Common ion effect

The common ion effect explains the precipitation of relatively soluble salts for which a solubility product cannot be defined.

The equilibrium existing in a saturated solution of a soluble electrolyte AB is

$$AB(s) \rightleftharpoons A^+(aq) + B^-(aq)$$

The solid AB is precipitated by the addition of a solution containing one of the common ions $A^+(aq)$ or $B^-(aq)$. The effect is explained by assuming that the common ion competes with the ions already present for water of hydration. This effectively increases the concentration of $A^+(aq)$ and $B^-(aq)$ ions and precipitation occurs. For example the passage of hydrogen chloride gas into a saturated solution of sodium chloride causes precipitation of the salt.

Effect of temperature, pressure and concentration changes on equilibrium

The values of the equilibrium constants K_c and K_p are unaffected by changes in pressure and concentration.

Both K_c and K_p are altered by changes in temperature according to the relationship

$$\log_e K = \frac{\Delta H^\ominus}{RT} + \text{constant}$$

where $\Delta H^\ominus$ is the standard enthalpy change for the reaction; R is the gas constant; T is the absolute temperature.
(This is derived from thermodynamics.)

∴ For endothermic reactions, K increases with temperature
For exothermic reactions, K decreases with temperature

Le Chatelier's Principle

Le Chatelier's Principle is used to predict qualitatively the effect of changes in temperature, pressure and concentration on a given equilibrium.

It states that if a system is in equilibrium and a change occurs in one of the equilibrium conditions, then the equilibrium readjusts itself so as to counteract the effect of the change.

For changes in concentration and pressure, the system readjusts itself so that the concentrations or partial pressures again satisfy the expressions for K_c and K_p. For changes in temperature, the system adjusts itself to the new value of K_c or K_p.

Note: The addition of a catalyst does not alter the composition of the equilibrium mixture nor the value of K_c or K_p. It increases the rate of both the forward and backward reactions so that equilibrium is reached more quickly.

(a) **Concentration** If a system is in equilibrium and the concentration of the reactants is increased, or the concentration of products decreased, the system will readjust itself so that there is a larger concentration of products in the new equilibrium mixture. If a system is in equilibrium and the concentration of reactants is decreased, or the concentration of products increased, the concentration of products in the new equilibrium mixture will be smaller, e.g.

$$H_2 + I_2 \rightleftharpoons 2HI.$$

If the concentration of iodine is increased more hydrogen iodide will be present in the new equilibrium mixture.

(b) **Pressure** Only equilibrium reactions involving gases are significantly affected by pressure changes. Gas phase equilibria are affected by pressure changes if the molecularity of the products is different from that of the reactants. An increase in pressure leads to an adjustment of the equilibrium in favour of the side of the reaction with smaller volume (i.e. lower molecularity) and vice versa, e.g.

$$N_2 + 3H_2 \rightleftharpoons 2NH_3$$

An increase in pressure results in a higher proportion of ammonia in the new equilibrium mixture as the forward reaction leads to a decrease in volume.

(c) **Temperature** A change in temperature results in a new equilibrium being established for which the equilibrium constant is different from that for the initial conditions.
Quantitatively, if the forward reaction is endothermic then an increase in temperature results in a higher concentration of products in the new equilibrium mixture and vice versa, e.g.

$$H_2O(g) + C(s) \rightleftharpoons CO(g) + H_2(g) \qquad \Delta H^{\ominus}_{298} = +130 \text{ kJ}$$

High temperature favours the yield of products.

Application of Le Chatelier's Principle

1 Haber Process

$$3H_2(g) + N_2(g) \rightleftharpoons 2NH_3(g) \qquad \Delta H^{\ominus}_{298} = -100 \text{ kJ mol}^{-1}$$

Theoretically a low temperature and high pressure would increase the yield of ammonia. In practice a pressure of about 350 atmospheres is used and a temperature of about 500°C (a lower temperature decreases the rate at which equilibrium is reached). A catalyst of iron oxides promoted with aluminium oxide and potassium hydroxide is used. About 30 % conversion occurs and after removal of the ammonia the gases are recycled.

2 Contact Process

$$2SO_2(g) + O_2(g) \rightleftharpoons SO_3(g)$$
$$\Delta H^{\ominus}_{298} = -94{\cdot}6 \text{ kJ mol}^{-1}$$

Theoretically, a high pressure and low temperature would increase the yield of sulphur(VI) oxide. However, only 2 atmospheres pressure and about 450°C are used as under these conditions both the yield and reaction rate are satisfactory. A catalyst of vanadium(V) oxide is used and a conversion rate of about 98 % is achieved. The sulphur(VI) oxide is dissolved in concentrated sulphuric acid as its reaction with water is very exothermic and an acid steam is formed.

Practice questions

In questions 1 to 11 give the letter of the correct answer.

1 In which of the following reactions will an increase in total pressure at constant temperature favour the formation of products?
A. $2C(s) + O_2(g) \rightleftharpoons 2CO(g)$
B. $H_2(g) + I_2(g) \rightleftharpoons 2HI(g)$
C. $2SO_3(g) \rightleftharpoons O_2(g) + 2SO_2(g)$
D. $2NH_3(g) \rightleftharpoons N_2(g) + 3H_2(g)$
E. $C_2H_4(g) + H_2(g) \rightleftharpoons C_2H_6(g)$

2 When 1 mole of hydrogen iodide is heated to 1000 K, 50% of the compound dissociates according to the equilibrium $2HI \rightleftharpoons H_2 + I_2$. Which of the following is the number of moles of gas in the equilibrium mixture?
A. 2 B. 1·5 C. 1·0 D. 0·75 E. 0·5

3 For the equilibrium $Ni(CO)_4(g) \rightleftharpoons Ni(s) + 4CO(g)$, appropriate units of K_p are
A. g atm^3 B. atm C. N^3 m^{-6} D. N m^{-2} E. mol atm^3

4 A saturated aqueous solution of a compound XY contains 6×10^{-3} g ions l^{-1} of X^{2+}(aq) at 298 K. Which of the following is the solubility product of XY at 298 K?
A. $1{\cdot}2 \times 10^{-2}$ B. $3{\cdot}6 \times 10^{-5}$ C. $3{\cdot}6 \times 10^{-6}$
D. $1{\cdot}2 \times 10^{-4}$ E. 9×10^{-6}

5 In the equilibrium $X_2 + Y_2 \rightleftharpoons 2XY$, $K = 20$, which of the following is the number of moles of X_2 in the equilibrium mixture if it contains 1 mole of Y_2 and 2 moles of XY?
A. 0·1 B. 0·2 C. 5 D. 10 E. 20

6 In the equilibrium $A + B \rightleftharpoons C + D$, the initial concentrations of both A and B are 1·6 $mol^{-1}\ l^{-1}$. At 298 K the equilibrium mixture contains 1·2 mol l^{-1} of C. Which of the following is the value of K_c for the equilibrium at 298 K?
A. 0·11 B. 1·77 C. 3 D. 5 E. 9

7 The equilibrium constants for the following are K_1 and K_2 respectively.

$$\tfrac{1}{2}N_2(g) + \tfrac{1}{2}O_2(g) \rightleftharpoons NO(g)$$
$$NO(g) + \tfrac{1}{2}O_2(g) \rightleftharpoons NO_2(g)$$

Which of the following is the equilibrium constant for the equation

$$NO_2(g) \rightleftharpoons \tfrac{1}{2}N_2(g) + O_2(g)?$$

A. $\dfrac{1}{K_1} \times \dfrac{1}{K_2}$ B. $K_1 \times K_2$ C. $K_1 + K_2$ D. $K_1 - K_2$
E. $\dfrac{K_2}{K_1}$

8 Which of the following favours the formation of $[Cu(H_2O)_4]^{2+}(aq)$ in the equilibrium

$$4Cl^-(aq) + [Cu(H_2O)_4]^{2+}(aq) \rightleftharpoons [CuCl_4]^{2-}(aq) + 4H_2O(l)$$

Addition of:
A. Water
B. Aqueous NaCl
C. A large excess of ammonia
D. Concentrated hydrochloric acid
E. Aqueous $CaCl_2$

9 Assuming the solubility of $PbCl_2$ is x mol l^{-1}, the value of the solubility product at the same temperature is
A. x^3 B. x^2 C. $x^2/4$ D. $4x^3$ E. $x^3/4$

10 Which of the following sets of units is correct for quoting the dissociation constant for the equilibrium $X_2Y(g) \rightleftharpoons 2X(g) + Y(g)$?
A. $mol^2\ l^{-2}$ B. mol^2 C. $mol\ l^{-1}$ D. mol^3 E. $mol^3\ l^{-3}$

11 The solubility of $CaCl_2$ is Y g l^{-1} and its relative molecular mass is Z. The solubility product of $CaCl_2$ is
A. $\dfrac{Y^2}{Z^2}$ B. $\dfrac{Y^3}{Z^3}$ C. $\dfrac{2Y^3}{Z^3}$ D. $\dfrac{4Y^3}{Z^3}$ F. $4Y^3Z^3$

12 State Le Chatelier's Principle.
A sample of dinitrogen textroxide has an average relative molecular mass of 52 at 85°C and 1 atm.

(a) Give an equation for the dissociation of dinitrogen textroxide.
(b) Calculate the degree of dissociation and the dissociation constant at 85°C and 1 atm.
(c) The dissociation is endothermic. Give with reasons the effect of increasing the temperature on the dissociation constant.

13 (a) Give the equation for the formation of ethyl ethanoate from ethanol and ethanoic acid. Name a catalyst for the reaction.
(b) If K_c for the equilibrium is 4 at 25°C, calculate the mass of ester formed if 2 moles of the acid and 2 moles of the alcohol are mixed and allowed to reach equilibrium.

14 For the equilibrium $2SO_2(g) + O_2(g) \rightleftharpoons 2SO_3(g)$ $\Delta H = -x$ kJ mol^{-1}
(a) Write an expression for K_p.
(b) State with reasons the effect on the equilibrium of
(i) increasing the pressure at constant temperature, and
(ii) increasing the temperature at constant pressure.
(c) Give the values of temperature and pressure used in the manufacture of SO_3.

15 Describe the Haber synthesis of ammonia and discuss the principles used for deciding optimum conditions.

Explain why a hot piece of platinum wire held above concentrated aqueous ammonia continues to glow even when no further heat is applied. Give the function of the platinum.

Ammonium hydrogensulphide NH_4HS dissociates into ammonia and hydrogen sulphide. Some of the solid salt is placed in an evacuated vessel at 25°C. What pressure is developed in the vessel? (K_p for NH_4HS at 25°C is $9{\cdot}56 \times 10^{-2}$ atm^2.)

7 Redox Equilibria

Oxidation and reduction

Oxidation is the loss of electrons, e.g.

$$Fe^{2+} \longrightarrow Fe^{3+} + e$$

Reduction is the gain of electrons, e.g.

$$Fe^{3+} + e \longrightarrow Fe^{2+}$$

An oxidizing agent or oxidant is a substance which brings about oxidation. It is therefore an acceptor of electrons and is itself reduced.

A reducing agent or reductant is a substance which brings about reduction. It is therefore a donor of electrons and is itself oxidized.

The equations describing separate oxidation and reduction reactions are called half equations and always include one or more electrons.

Redox reactions

A redox reaction is a reaction in which both oxidation and reduction occur. It therefore involves the transfer of electrons, the number of electrons lost in oxidation being equal to the number gained in reduction, e.g.

$$Cu^{2+}(aq) + Zn(s) \longrightarrow Zn^{2+}(aq) + Cu(s)$$

Reduction $\quad Cu^{2+}(aq) + 2e \longrightarrow Cu(s)$

Oxidation $\quad Zn(s) \longrightarrow Zn^{2+}(aq) + 2e$

Oxidation number (O.N.) or oxidation state

The O.N. of an element in an ionic compound is the charge on its ion in that compound. The O.N. of an element in a covalent compound is the charge the atom would have if the bonding were to be ionic taking into account the relative electronegativities of the species present, e.g.

	Na^+	Cl^-	H	Cl	Cl	F	C	Cl_4
O.N.	+1	−1	+1	−1	+1	−1	+4	−1

Rules

1 The algebraic sum of the oxidation numbers in a covalent compound is zero. The algebraic sum of the oxidation numbers in an ion equals the charge on the ion, e.g.

H_2O Sum of O.N. $= 2(+1) + (-2) = 0$
NH_4^+ Sum of O.N. $= (-3) + 4(+1) = +1$

2 Elements in their uncombined states have an O.N. of 0. For example in Cl_2, O.N. of Cl = 0.

3 Fluorine has an O.N. of −1 as it is the most electronegative element.

4 Oxygen has an O.N. of −2 except in (a) F_2O, where it has a value of +2, (b) peroxides, such as H_2O_2, where it has a value of −1.

5 Hydrogen has an O.N. of +1 except in the metallic hydrides such as NaH where it has a value of −1.

6 Some elements have more than one O.N. When present in a particular compound the element's O.N. is calculated by using the known values for the other elements present, e.g.

H_2SO_4 Sum of O.N. $= 2(+1) +$ O.N. of S $+ 4(-2) = 0$
$\therefore$ O.N. of S $= 8 - 2 = +6$

Na_2SO_3 Sum of O.N. $= 2(+1) +$ O.N. of S $+ 3(-2) = 0$
$\therefore$ O.N. of S $= 6 - 2 = +4$

Transition metals usually have more than one O.N. For example Mn in MnO_2, MnO_4^{2-}, MnO_4^- has the values +4, +6 and +7 respectively.

7 When an element is oxidized, its O.N. is increased and when reduced it is decreased. The change in O.N. is the number of electrons in the half equation, e.g.

$$\underset{2(+3)}{2Fe^{3+}(aq)} + \underset{-2}{S^{2-}(aq)} \longrightarrow \underset{2(+2)}{2Fe^{2+}(aq)} + \underset{0}{S(s)}$$

Reduction $\underset{+3}{Fe^{3+}} + e \longrightarrow \underset{+2}{Fe^{2+}}$ Oxidation $\underset{-2}{S^{2-}} \longrightarrow \underset{0}{S} + 2e$

8 O.N.s provide the basis for the stock system of naming compounds, e.g. $KMnO_4$ is potassium manganate(VII) as the O.N. of Mn is +7.

Writing equations for redox reactions

1 Give the oxidation and reduction half equations by:
(a) Writing down the species reduced or oxidized and the product formed.
(b) Balancing with respect to atoms.
If the reaction is in acidic conditions balance with $H^+(aq)$ and $H_2O(l)$.
If the reaction is in alkaline conditions balance with $OH^-(aq)$ and $H_2O(l)$.
(c) Balancing with respect to charges by the addition of the appropriate number of electrons.

2 Adjust the half equations so that the number of electrons lost equals the number gained.

3 Add the half equations together, cancelling where appropriate.

Example
Give the ionic equation for the oxidation of iron(II) sulphate in acidic solution by potassium dichromate(VI) solution.

Oxidation $Fe^{2+}(aq) \longrightarrow Fe^{3+}(aq) + e$

Reduction $Cr_2O_7^{2-}(aq) + 14H^+(aq) + 6e \longrightarrow 2Cr^{3+}(aq) + 7H_2O(l)$

$\therefore 6Fe^{2+}(aq) + CrO_6^{2-}(aq) + 14H^+(aq) \longrightarrow 6Fe^{3+}(aq) + 2Cr^{3+}(aq) + 7H_2O(l)$

Common oxidizing agents

Reagent	**Half equation**
$KMnO_4$ in acid solution	$MnO_4^-(aq) + 8H^+(aq) + 5e \longrightarrow Mn^{2+}(aq) + 4H_2O(l)$
$KMnO_4$ in alkaline solution	$MnO_4^-(aq) + 2H_2O(l) + 3e \longrightarrow MnO_2(s) + 4OH^-(aq)$
$K_2Cr_2O_7$ in acid solution	$Cr_2O_7^{2-}(aq) + 14H^+(aq) + 6e \longrightarrow 2Cr^{3+}(aq) + 7H_2O(l)$
Dilute HNO_3	$NO_3^-(aq) + 4H^+(aq) + 3e \longrightarrow NO(g) + 2H_2O(l)$
Concentrated HNO_3	$NO_3^-(aq) + 2H^+(aq) + e \longrightarrow NO_2(g) + H_2O(l)$
Concentrated H_2SO_4	$SO_4^{2-}(aq) + 4H^+(aq) + 2e \longrightarrow SO_2(g) + 2H_2O(l)$
MnO_2	$MnO_2(s) + 4H^+(aq) + 2e \longrightarrow Mn^{2+}(aq) + 2H_2O(l)$
Cl_2	$Cl_2(g) + 2e \longrightarrow 2Cl^-(aq)$

Common reducing agents

Reagent	**Half equation**
Fe^{2+} salts	$Fe^{2+}(aq) \longrightarrow Fe^{3+}(aq) + e$
Ethanedioates	$C_2O_4^{2-}(aq) \longrightarrow 2CO_2(g) + 2e$
Sulphites and moist SO_2	$SO_3^{2-}(aq) + H_2O \longrightarrow SO_4^{2-}(aq) + 2H^+(aq) + 2e$

H_2S	$S^{2-}(aq) \longrightarrow S(s) + 2e$
Iodides	$2I^-(aq) \longrightarrow I_2(s) + 2e$
Metals, e.g. Zn	$Zn(s) \longrightarrow Zn^{2+}(aq) + 2e$
H_2	$H_2 \longrightarrow 2H^+(aq) + 2e$

Electrode potentials

When a metal is placed in an aqueous solution of its ions the following equilibrium is set up.

$$M(s) \rightleftharpoons M^{Z+}(aq) + Ze$$

If the equilibrium is well over to the right then the metal will be negatively charged with respect to the solution and vice versa. An electrical potential is therefore set up between the metal and its ions in solution. Its value depends on the position of equilibrium and so varies from metal to metal.

It is not possible to measure directly the electrode potential of a single metal/metal ion system but the difference in potential between two different metal/metal ion systems can be found by determining the e.m.f. of the appropriate cell. In practice the metal/metal ion system used for comparison is the standard hydrogen electrode and it itself is given an electrode potential of zero.

The standard hydrogen electrode

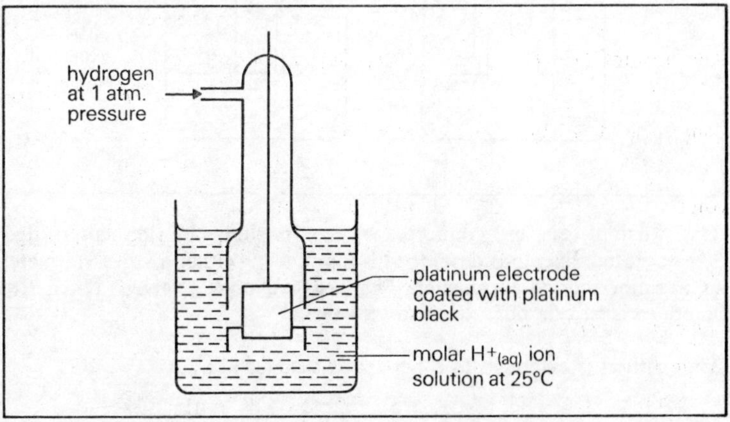

The hydrogen is absorbed on the platinum electrode which catalyses the setting up of the following equilibrium

$$2H^+(aq) + 2e \rightleftharpoons H_2(g)$$

The electrode potential varies with temperature and ionic concentration so a 1 *M* solution of H^+(aq) ions at 25°C is used.

Standard electrode potential $E^\ominus$

The standard electrode potential of a metal is the potential difference set up between a standard hydrogen electrode and a half-cell containing the metal in contact with a 1 molar solution of its ions at 25°C, i.e. it is the e.m.f. of the cell

$$Pt[H_2(g)]|2H^+(aq)\vdots M^{z+}(aq)|M(s)$$

The sign of the standard electrode potential of the metal is the polarity of its electrode in the following cell.

Experimental determination of standard electrode potentials

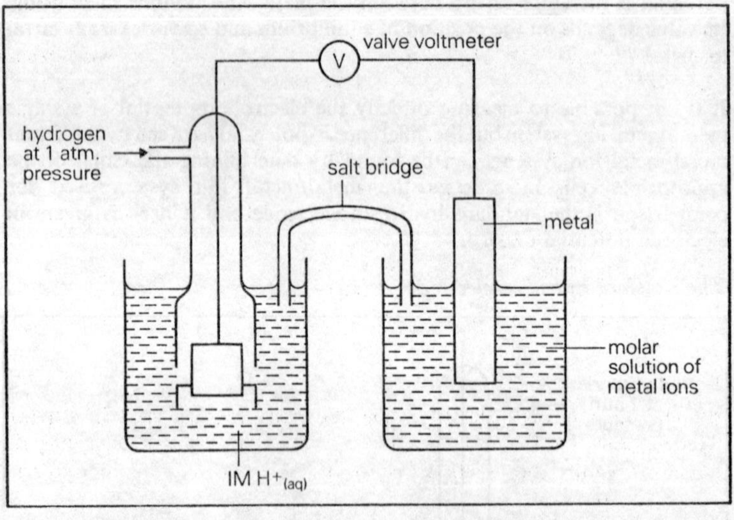

The two half cells are connected by a potassium chloride salt bridge. The potential difference developed is found using either a valve voltmeter or a calibrated potentiometer. The negative pole is given a negative standard electrode potential and vice versa.

The method is extended to other redox systems such as

$$Fe^{3+}(aq) + e \rightleftharpoons Fe^{2+}(aq)$$

$$I_2(aq) + 2e \rightleftharpoons 2I^-(aq)$$

The cell notation is reduced form | oxidized form ⁝ oxidized form | reduced form.

Redox potential series

The redox potential series is an arrangement of reduction systems in ascending order of their standard electrode potentials.

Half reaction	$E^{\ominus}$ in volts
$Cs^+ + e \longrightarrow Cs$	−3·04
$Li^+ + e \longrightarrow Li$	−3·02
$K^+ + e \longrightarrow K$	−2·92
$Ca^{2+} + 2e \longrightarrow Ca$	−2·76
$Na^+ + e \longrightarrow Na$	−2·71
$Mg^{2+} + 2e \longrightarrow Mg$	−2·40
$Al^{3+} + 3e \longrightarrow Al$	−1·66
$Zn^{2+} + 2e \longrightarrow Zn$	−0·76
$S + 2e \longrightarrow S^{2-}$	−0·51
$Fe^{2+} + 2e \longrightarrow Fe$	−0·44
$Pb^{2+} + 2e \longrightarrow Pb$	−0·13
$2H^+ + 2e \longrightarrow H_2$	0·00
$SO_4^{2-} + 2H^+ + 2e \longrightarrow SO_3^{2-} + H_2O$	+0·17
$Cu^{2+} + 2e \longrightarrow Cu$	+0·35
$I_2 + 2e \longrightarrow 2I^-$	+0·58
$Ag^+ + e \longrightarrow Ag$	+0·81
$Cr_2O_7^{2-} + 14H^+ + 6e \longrightarrow 2Cr^{3+} + 7H_2O$	+1·30
$Cl_2 + 2e \longrightarrow 2Cl^-$	+1·36
$MnO_4^- + 8H^+ + 5e \longrightarrow Mn^{2+} + 4H_2O$	+1·50
$F_2 + 2e \longrightarrow 2F^-$	+2·85

↓ (left side, downwards) increasing strength of oxidant, i.e. increasing ease of reduction

↑ (right side, upwards) increasing strength of reductant, i.e. increasing ease of oxidation

Points

1 The more positive the $E^{\ominus}$ value, the more readily the half reaction occurs in the direction indicated. The more negative the value, the more readily the reaction occurs in the opposite direction to that indicated.

2 F_2 is the strongest oxidizing agent and Cs^+ ion the weakest. Oxidizing strength increases regularly from Cs^+ to F_2.

3 Cs is the strongest reducing agent and F^- ion the weakest. Reducing strength increases regularly from F^- to Cs.

Uses of the redox potential series

1 **Prediction of metal and non-metal reactivity**

In general the series lists the metals in decreasing order of reactivity and non-metals in increasing order. However, exceptions occur, e.g. aluminium is much less reactive than its $E^{\ominus}$ value indicates as it is covered by a protective oxide layer.

2 Calculation of the e.m.f. of a cell and $E^\ominus$ values

Note: In general the e.m.f. of a cell = standard electrode potential of right-hand cell − standard electrode potential of left-hand cell

Examples

(a) The Daniell cell

The Daniell cell may be represented by

$$Cu(s) \mid Cu^{2+}(aq) \,\vdots\, Zn^{2+}(aq) \mid Zn(s)$$

$$\begin{aligned} \text{e.m.f. of cell} &= E^\ominus_{Zn^{2+}|Zn} - E^\ominus_{Cu^{2+}|Cu} \\ &= -0{\cdot}76 \quad - (+0{\cdot}35)\ V \\ &= -1{\cdot}11\ V \end{aligned}$$

The zinc electrode is therefore negative.
Since the e.m.f. is negative, the actual reaction which occurs is the reverse of that suggested above in the cell representation, i.e. reaction occurring is

$$Zn(s) + Cu^{2+}(aq) \longrightarrow Cu(s) + Zn^{2+}(aq)$$

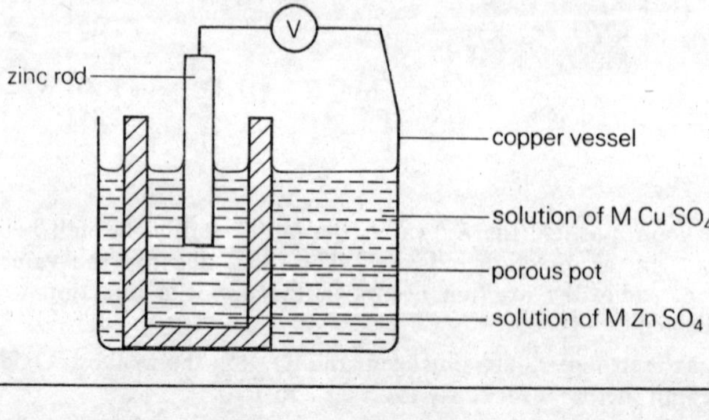

The cell is set up as shown. When the two electrodes are connected a current flows from the copper to the zinc and electrons flow in the opposite direction, i.e.

$$\begin{aligned} Zn &\longrightarrow Zn^{2+} + 2e \qquad \text{Oxidation} \\ Cu^{2} + 2e &\longrightarrow Cu \qquad\qquad \text{Reduction} \end{aligned}$$

The potential difference between the two electrodes increases with decreasing resistance being a maximum when no current flows. It is then called the e.m.f. of the cell and is found to be 1·11 V.

(b) Calculate the value of $E^{\ominus}$ $Ag^+|Ag$ given that the e.m.f. of the cell $Ag(s)|Ag^+(aq)\vdots Zn^{2+}(aq)|Zn(s)$ is $-1{\cdot}57$ V and $E^{\ominus}_{Zn^{2+}|Zn}$ is $-0{\cdot}76$ V.
Give the cell reaction.

$$\text{e.m.f. of cell} = E^{\ominus}_{Zn^{2+}|Zn} - E^{\ominus}_{Ag^+|Ag}$$

$$\therefore \quad -1.57 = -0{\cdot}76 - E^{\ominus}_{Ag^+|Ag}$$

$$\therefore \quad E^{\ominus}_{Ag^+|Ag} = -0{\cdot}76 + 1{\cdot}57$$

$$= +0{\cdot}81 \text{ V}$$

Since the e.m.f. of the cell is negative the reaction which occurs is

$$2Ag^+(aq) + Zn(s) \longrightarrow 2Ag(s) + Zn^{2+}(aq)$$

3 Prediction of the feasibility of redox reactions

In general the 'anticlockwise rule' is applied to predict the direction of change between 2 half reactions, e.g.

$$Mg^{2+}(aq) + 2e \longrightarrow Mg(s) \qquad E^{\ominus} = -2{\cdot}40 \text{ V}$$

$$Pb^{2+}(aq) + 2e \longrightarrow Pb(s) \qquad E^{\ominus} = -0{\cdot}13 \text{ V}$$

The half reactions are listed as shown in ascending order of $E^{\ominus}$ values. By the 'anticlockwise rule' the second reaction should go in the direction indicated and the first in the opposite direction.
∴ The reaction expected to occur is

$$Mg(s) + Pb^{2+}(aq) \longrightarrow Pb(s) + Mg^{2+}(aq).$$

This is confirmed by calculating the e.m.f. of the cell which could be set up.
For the cell $Mg(s)|Mg^{2+}(aq)\vdots Pb^{2+}(aq)|Pb(s)$ the e.m.f. $= -0{\cdot}13 - (-2{\cdot}40) = +2{\cdot}27$ V.
If the e.m.f. is positive then theoretically the reaction should go in the direction indicated. (In practice the e.m.f. value should be greater than 0·4 V.)
In general metals higher in the redox potential series displace metals below them from solutions of their ions.
Non-metals lower in the redox potential series displace non-metals above them from solutions of their ions.

Effect of concentration and temperature on values of electrode potentials

1 Concentration

$$M^{z+}(aq) + Ze \rightleftharpoons M(s)$$

By Le Chatelier's Principle, if the metal ion concentration is increased then the equilibrium will be displaced to the right. The electrode potential will therefore become more positive (i.e. less negative). If the metal ion concentration is decreased then the electrode potential becomes less positive.

2 Temperature

The electrode potential becomes more positive as the temperature is raised and vice versa.

The effect of concentration and temperature on electrode potentials is given by the Nernst equation.

$$E = E^{\ominus} + \frac{RT}{ZF} \log_e \frac{[\text{oxidized form}]}{[\text{reduced form}]}$$

where E is the electrode potential in V;
$E^{\ominus}$ is the standard electrode potential in V;
R is the gas constant (8·314 J K^{-1} mol^{-1});
T is the temperature in K;
Z is the change in oxidation number between the oxidized and reduced forms;
F is the Faraday (96500 coulombs)

The equation applies to dilute solutions only. For metal/metal ion systems, the reduced form is the metal and this term is omitted from the equation.

Practice questions

In questions 1 to 10 give the letter of the correct answer.

1 Which of the following statements is correct about the operation of the cell $Ag(s) | Ag^+(aq) ⁞ Zn^{2+}(aq) | Zn(s)$ when the electrodes are connected via an external circuit?

A. Electrons flow in the external circuit from the silver to zinc electrode.
B. Zinc is deposited on the zinc electrode.
C. Zinc atoms undergo reduction.
D. The zinc electrode forms the negative pole of the cell.
E. The silver electrode decreases in mass.

2 $E^{\ominus}$ values for the systems $Fe^{2+}(aq) | Fe(s)$ and $Cu^{2+}(aq) | Cu(s)$ are −0·4 V and +0·4 V respectively. Which of the following is the standard e.m.f. in V of the cell $Fe(s) | FeSO_4(aq) ⁞ CuSO_4(aq) | Cu(s)$?

A. −0·8 B. −0·4 C. 0 D. 0·4 E. 0·8

3 In which of the following does the metal undergo a change in its oxidation state?

A. $Ni + 4Co \longrightarrow Ni(CO)_4$
B. $Cr_2O_7^{2-} + H_2O \rightleftharpoons 2CrO_4^{2-} + 2H^+$
C. $[Co(H_2O)_6]^{2+} + 4Cl^- \longrightarrow [CoCl_4]^{2-} + 6H_2O$
D. $2MnO_4^{2-} + Cl_2 \longrightarrow 2MnO_4^- + 2Cl^-$
E. $[Fe(H_2O)_6]^{2+} + 6CN^- \longrightarrow [Fe(CN)_6]^{4-} + 6H_2O$

Questions 4 to 7
Given the following $E^{\ominus}$ values:

$$Mg^{2+}(aq)\,|\,Mg(s) = -2{\cdot}37\ V,$$
$$Al^{3+}(aq)\,|\,Al(s) = 1{\cdot}66\ V,$$
$$Zn^{2+}(aq)\,|\,Zn(s) = -0{\cdot}76\ V,$$
$$Ag^{+}(aq)\,|\,Ag(s) = +0{\cdot}81\ V,$$

choose from the list A to E the cell reactions which correctly answer questions 4 to 7.

A. $Mg(s) + Zn^{2+}(aq) \longrightarrow Mg^{2+}(aq) + Zn(s)$
B. $Zn(s) + 2Ag^{+}(aq) \longrightarrow 2Ag(s) + Zn^{2+}(aq)$
C. $2Al(s) + 3Mg^{2+}(aq) \longrightarrow 3Mg(s) + 2Al^{3+}(aq)$
D. $2Al(s) + 3Zn^{2+}(aq) \longrightarrow 3Zn(s) + 2Al^{3+}(aq)$
E. $Mg(s) + 2Ag^{+}(aq) \longrightarrow Mg^{2+}(aq) + 2Ag(s)$

4 Which reaction is a basis of a cell with an e.m.f. of 1·57 V?

5 Which reaction is the basis of the cell with highest e.m.f. value?

6 Which reaction does not occur in the direction written?

7 Which reaction is the basis of the cell with the smallest positive e.m.f. value?

Questions 8 to 10
Questions 8 to 10 refer to the following half reactions

$$MnO_4^{-} + 8H^{+} + 5e \longrightarrow Mn^{2+} + 4H_2O,$$
$$Fe^{2+} \longrightarrow Fe^{3+} + e,$$
$$C_2O_4^{2-} \longrightarrow 2CO_2 + 2e$$

8 The number of moles of iron(II) ion oxidized by 1 mole of MnO_4^{-} ion is
A. $\frac{1}{5}$ B. $\frac{2}{5}$ C. 1 D. $\frac{3}{2}$ E. 5

9 The number of moles of ethanedioate ion oxidized by 1 mole of MnO_4^{-} ion is
A. $\frac{1}{5}$ B. $\frac{2}{5}$ C. 1 D. $\frac{5}{2}$ E. 5

10 The number of moles of $NaHC_2O_4{\cdot}H_2C_2O_4{\cdot}2H_2O$ oxidized by 1 mole of MnO_4^{-} ion is
A. $\frac{2}{5}$ B. $\frac{4}{5}$ C. 1 D. $\frac{5}{4}$ E. $\frac{5}{2}$

11 (a) For the reaction

$$2Cu^{2+} + 2Cl^{-} + 2H_2O + SO_2 \longrightarrow SO_4^{2-} + 4H^{+} + 2CuCl$$

name the species oxidized and reduced and give half reactions for the oxidation and reduction.

(b) What is meant by the term 'disproportionation' and give one example.

(c) In volumetric analysis, cerium(IV) solutions can be used to estimate ethanedioate solutions. The half equations are

$$Ce^{4} + e \longrightarrow Ce^{3+},$$

$$C_2O_4{}^{2-} \longrightarrow 2CO_2 + 2e.$$

Calculate the volume of 0·05 *M* cerium(IV) solution used against a solution containing 0·0025 moles of ethanedioate ions.

12 (a) The following circuit is set up.

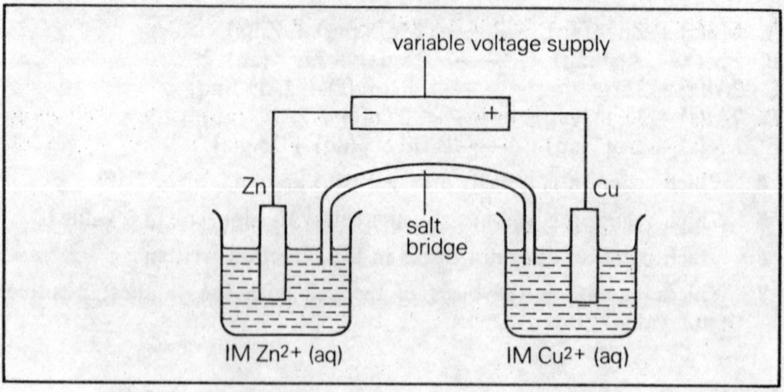

The $E^{\ominus}$ values for $Cu^{2+} \mid Cu$ and $Zn^{2+} \mid Zn$ are $+0{\cdot}34$ and $-0{\cdot}76$ V respectively.

Write down the equation for the reaction occurring in the cell and the direction of electron flow in the wire if the variable voltage is set at (i) 0·9 V and (ii) 1·3 V.

(b) Describe, with the aid of a diagram, an experiment to measure the standard electrode potential of silver and write an equation representing the cell reaction.

8 Acids and Bases

Brønsted–Lowry theory

This theory defines:

(a) **An acid as a substance which dissociates to produce protons when it is dissolved in certain solvents.**

(b) **A base as a substance which combines with protons.**

Examples

(i) When hydrogen chloride is dissolved in water the following equilibrium is set up in which the water acts as a base by accepting protons.

$$\underset{\text{acid 1}}{HCl(aq)} + \underset{\text{base 2}}{H_2O(l)} \rightleftharpoons \underset{\text{acid 2}}{H_3O^+(aq)} + \underset{\text{base 1}}{Cl^-(aq)}$$

Acid 1 and base 1 are known as a conjugate pair as are acid 2 and base 2.
The equilibrium is well over to the right as HCl loses protons more readily than H_3O^+ and H_2O accepts protons more readily than Cl^-. In this reaction, therefore, HCl is said to act as a stronger acid than H_3O^+ and water as a stronger base than Cl^-.

(ii) When ammonia is dissolved in water the following equilibrium is set up in which the water acts as an acid by donating protons.

$$\underset{\text{base 1}}{NH_3(aq)} + \underset{\text{acid 2}}{H_2O(l)} \rightleftharpoons \underset{\text{acid 1}}{NH_4^+(aq)} + \underset{\text{base 2}}{OH^-(aq)}$$

In the solution there is a higher concentration of NH_3 than of NH_4^+ indicating that NH_4^+ loses protons more readily than H_2O and that OH^- accepts protons more readily than NH_3.

The strength of acids and bases depends on the solvent used. Thus, when hydrogen chloride is dissolved in water it acts as a strong acid as the equilibrium is well over to the right. The hydrogen chloride is therefore almost completely dissociated. However, when hydrogen chloride is dissolved in glacial ethanoic acid it is much weaker as the ethanoic acid does not accept protons as readily as water.

$$HCl + CH_3COOH \rightleftharpoons CH_3COOH_2^+ + Cl^-$$

The equilibrium is therefore further to the left and the hydrogen chloride is only partially dissociated.

Lewis theory

This theory defines:

(a) **An acid as a substance which accepts a lone pair of electrons in forming a co-ordinate bond.**

(b) **A base as a substance which donates a lone pair of electrons in forming a co-ordinate bond.**

e.g. (i)

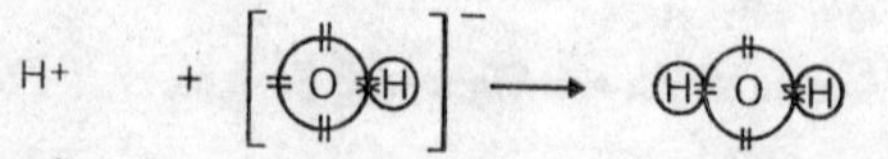

(ii)

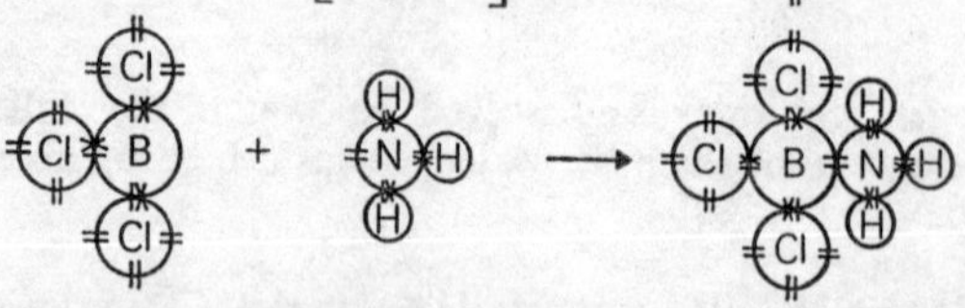

ammonia boron trichloride adduct

Ionic equilibria

The equilibrium law is applicable to only dilute solutions of ions. For more concentrated solutions, the ions interfere and the concentration terms are invalid.

Equilibrium constant for the ionization of water

The following equilibrium exists in water

$$H_2O(l) \rightleftharpoons H^+(aq) + OH^-(aq)$$

Applying the equilibrium law

$$K_c = \frac{[H^+(aq)][OH^-(aq)]}{[H_2O(l)]}$$

The concentration of unionized water is considered to be constant

$$\therefore [H^+(aq)][OH^-(aq)] = \text{constant} \times K_c = K_w$$

K_w is known as the ionic product of water and is constant at constant temperature. K_w increases with an increase in temperature as the ionization of water is endothermic.

At 298 K $\qquad K_w = 10^{-14}$

At 333 K $\qquad K_w = 9{\cdot}5 \times 10^{-14}$

In neutral solutions $[H^+(aq)] = [OH(aq)]$

$\therefore$ at 298 K $\qquad [H^+(aq)] = 10^{-7}$ g ions l^{-1}

If $[H^+(aq)] > 10^{-7}$ g ion l^{-1}, the solution is acidic.
If $[H^+(aq)] < 10^{-7}$ g ion l^{-1}, the solution is alkaline.

pH

The pH scale is a convenient scale for describing the degree of acidity or alkalinity of a solution. Sørensen defined pH as

$$pH = -\log_{10} [H^+(aq)] \text{ or } [H^+(aq)] = 10^{-pH}.$$

Since at 298 K

$$[H^+(aq)][OH^-(aq)] = 10^{-14}$$

$$-\log_{10} [H^+(aq)] - \log_{10} [OH^-(aq)] = +14.$$

But pOH is defined as $-\log_{10} [OH^-(aq)]$

$$\therefore pH + pOH = 14$$

Examples

(i) Find the pH and pOH of a solution with an $[H^+(aq)]$ of $5{\cdot}5 \times 10^{-5}$ g ions l^{-1}.

$$\begin{aligned} pH &= -\log_{10} [H^+(aq)] \\ &= -\log_{10} [5{\cdot}5 \times 10^{-5}] \\ &= -[\log_{10} 5{\cdot}5 + \log_{10} 10^{-5}] \\ &= -[0{\cdot}740 - 5] \end{aligned}$$

$$\therefore pH = 4{\cdot}26 \text{ (corr. to 3 sig. figs.).}$$

$$\begin{aligned} pOH &= 14 - 4{\cdot}26 \\ &= 9{\cdot}74 \text{ (corr. to 3 sig. figs.).} \end{aligned}$$

(ii) Find $[H^+(aq)]$ of a solution of pH 5·312

$$\begin{aligned} \log_{10} [H^+(aq)] &= -5{\cdot}312 \\ &= -6 + 0{\cdot}688 \end{aligned}$$

(It is impossible to antilog a negative number written in this form.)

$$\therefore [H^+(aq)] = 10^{-6} \times 4{\cdot}875 \text{ g ions } l^{-1}$$

∴ The hydrogen ion concentration of the solution is $4{\cdot}88 \times 10^{-6}$ g ions l^{-1} (corr. to 3 sig. figs.).

The pH scale is usually regarded as extending from 0 to 14 but negative pH values and values above 14 are possible. For example,

in 2 *M* HCl $[H^+(aq)] = 2$ g ions l^{-1} $\quad\therefore pH = -0{\cdot}301$.

However, the actual pH value will be larger as the solution is concentrated, the ions interfere and the concentration term is invalid.

$[H^+(aq)]$ g ions l^{-1}	10^{-1}	← increasing acid strength	10^{-7}	→	10^{-14}
pH	1		7		14
Type of solution	strong acid		neutral		strong alkaline
Example	1 *M* HCl		water		1 *M* NaOH

Strong acids and strong bases

These ionize completely in water to form solutions containing a high concentration of $H^+(aq)$ and $OH^-(aq)$ ions respectively

Strong acids, e.g. solutions of HCl, H_2SO_4, HNO_3 in water.

Strong bases, e.g. solutions of NaOH, KOH in water.

The contribution to the ionic concentrations of $H^+(aq)$ and $OH^-(aq)$ ions from the dissociation of the water is negligible and is ignored in pH calculations.

Examples

(i) Find the pH of 0·01 *M* H_2SO_4.

Assuming complete dissociation

$$[H^+(aq)] = 2 \times 10^{-2} \text{ g ions l}^{-1}$$
$$\therefore pH = -\log(2 \times 10^{-2})$$
$$= -[\log 2 + \log 10^{-2}]$$
$$= -[0{\cdot}3010 - 2]$$
$$= 1{\cdot}699$$

$\therefore$ pH of 0·01 *M* H_2SO_4 is 1·70 (corr. to 3 sig. figs.).

(ii) Find the pH of 0·05 *M* NaOH.

Assuming complete dissociation

$$[OH^-(aq)] = 5 \times 10^{-2} \text{ g ion l}^{-1}$$
$$pOH = -\log(5 \times 10^{-2})$$
$$= -[\log 5 + \log 10^{-2}]$$
$$= -[0{\cdot}6990 - 2]$$
$$= 1{\cdot}301$$
$$\therefore pH = 14 - pOH$$
$$= 14 - 1{\cdot}301$$
$$= 12{\cdot}699$$

$\therefore$ The pH of 0·05 *M* NaOH solution is 12·7 (corr. to 3 sig. figs.).

Weak acids and weak bases

Weak acids and weak bases do not ionize completely in solution. The pH of a weak acid solution is always higher than that of a strong acid of equal concentration, providing that the acids have the same basicity.

(a) **Weak acids,** e.g. ethanoic acid, phenol, carbonic acid
Consider the weak acid HA in solution in water. The following equilibrium is set up.

$$HA(aq) + H_2O(l) \rightleftharpoons H_3O^+(aq) + A^-(aq)$$

If the solution is dilute the equilibrium law can be applied.

$$K_c = \frac{[H_3O^+(aq)][A^-(aq)]}{[HA(aq)][H_2O(l)]}$$

The water concentration is considered to be constant.

$$\therefore\ K_c[H_2O(l)] = \frac{[H_3O^+(aq)][A^-(aq)]}{[HA(aq)]} = K_a$$

K_a is known as the dissociation constant of the acid and is constant at constant temperature. Values are usually quoted at 25°C.
The numerical value of K_a is usually small so pK_a values are used where

$$pK_a = -\log_{10} K_a.$$

Note that the smaller the pK_a value the stronger the acid. K_a does not vary with dilution unlike the degree of dissociation, α, of acids which increases with dilution.

Examples

(i) Calculate the pH of a 0·1 M solution of ethanoic acid if K_a for ethanoic acid is $1{\cdot}8 \times 10^{-5}$ mol l^{-1}.

$$K_a = 1{\cdot}8 \times 10^{-5} = \frac{[CH_3COO^-(aq)][H^+(aq)]}{[CH_3COOH(aq)]}$$

Ignoring the contribution to $[H^+]$ from the dissociation of the water

$$[H^+(aq)] = [CH_3COO^-(aq)].$$

Assuming the ethanoic acid is only slightly dissociated then

$$[CH_3COOH] = 0{\cdot}1 \text{ mol } l^{-1}$$

$$\therefore\ \frac{[H^+(aq)]^2}{0{\cdot}1} = 1{\cdot}8 \times 10^{-5}$$

$$\therefore\ [H^+(aq)] = \sqrt{1{\cdot}8 \times 10^{-6}}$$

$$\therefore\ [H^+(aq)] = 1{\cdot}34 \times 10^{-3}$$

$$\therefore\ pH = -\log_{10}(1{\cdot}34 \times 10^{-3})$$

$$= -[\log_{10} 1{\cdot}34 + \log_{10} 10^{-3}] = 2{\cdot}87$$

$\therefore$ The pH of a 0·1 M solution of ethanoic acid is 2·87 (corr. to 3 sig. figs.).

(ii) Calculate the pK_a of ethanoic acid if K_a of ethanoic acid = $1{\cdot}8 \times 10^{-5}$ mol l^{-1}.

$$\begin{aligned} pK_a &= -\log_{10}(1{\cdot}8 \times 10^{-5}) \\ &= -[\log 1{\cdot}8 + \log 10^{-5}] \\ &= -[0{\cdot}255 - 5] \\ &= 4{\cdot}75 \end{aligned}$$

∴ The pK_a of ethanoic acid is 4·75 (corr. to 3 sig. figs.).

(iii) Calculate the K_a of a methanoic acid solution if its pH is 2·9 and its concentration 0·01 *M*.

$$K_a = \frac{[HCOO^-][H^+]}{[HCOOH]}$$

Ignoring the H^+(aq) ions from the dissociation of water.

$$\begin{aligned} [H^+(aq)] &= [HCOO^-(aq)] \\ pH &= -\log_{10}[H^+(aq)] \\ &= 2{\cdot}9 \\ \therefore \log[H^+(aq)] &= -2{\cdot}9 \\ &= -3 + 0{\cdot}1 \\ \therefore [H^+(aq)] &= 1{\cdot}26 \times 10^{-3} \text{ g ions l}^{-1} \\ \therefore [HCOOH] &= 0{\cdot}01 - 1{\cdot}26 \times 10^{-3} \\ &= 10 \times 10^{-3} - 1{\cdot}26 \times 10^{-3} \text{ mol l}^{-1} \\ &= 8{\cdot}74 \times 10^{-3} \text{ mol l}^{-1} \\ \therefore K_a &= \frac{(1{\cdot}26 \times 10^{-3})^2}{8{\cdot}74 \times 10^{-3}} \\ &= 1{\cdot}82 \times 10^{-4} \text{ mol l}^{-1} \end{aligned}$$

∴ K_a for methanoic acid is $1{\cdot}82 \times 10^{-4}$ mol l^{-1} (corr. to 3 sig. figs.).

(b) **Weak bases,** e.g. ammonia
Consider the weak base A^- in solution in water in equilibrium with its conjugate acid HA.

$$A^-(aq) + H_2O(l) \rightleftharpoons HA(aq) + OH^-(aq)$$

Applying the equilibrium law

$$K_c = \frac{[HA(aq)][OH^-(aq)]}{[A^-(aq)][H_2O(l)]}$$

The water concentration is considered to be constant

$$\therefore\ K_c[H_2O(l)] = \frac{[HA(aq)][OH^-(aq)]}{[A^-(aq)]} = K_b$$

K_b is known as the basicity constant of the base and is constant at constant temperature. Values are quoted at 25°C.
Since K_b is often small, pK_b values are quoted where

$$pK_b = -\log_{10} K_b.$$

For conjugate acids and bases

$$K_a \times K_b = \frac{[H^+(aq)][A^-(aq)]}{[HA(aq)]} \times \frac{[HA(aq)][OH^-(aq)]}{[A^-(aq)]}$$

$$= [H^+(aq)][OH^-(aq)]$$

$$= K_w$$

At 25°C, $K_w = 10^{-14}$

$\therefore\ K_a \times K_b = 10^{-14}$

or $pK_a + pK_b = 14$

Examples

(i) Calculate the pH of a 0·01 M solution of methylamine if K_b for methylamine is $4{\cdot}5 \times 10^{-4}$ mol l^{-1}.

$$CH_3NH_2(aq) + H_2O(l) \rightleftharpoons CH_3NH_3^+(aq) + OH^-(aq)$$

$$K_b = \frac{[CH_3NH_3^+(aq)][OH^-(aq)]}{[CH_3NH_2(aq)]}$$

Ignoring the $[OH^-(aq)]$ from the dissociation of water,

$$[CH_3NH_3^+(aq)] = [OH^-(aq)].$$

Assuming the base is very weak then

$$CH_3NH_2(aq) \simeq 10^{-2}\ \text{mol l}^{-1}$$

$$\therefore\ \frac{[OH^-(aq)]^2}{10^{-2}} = 4{\cdot}5 \times 10^{-4}$$

$$\therefore\ [OH^-(aq)] = \sqrt{4{\cdot}5 \times 10^{-6}}$$

$$= 2{\cdot}12 \times 10^{-3}$$

$$\therefore\ [H^+(aq)] = \frac{10^{-14}}{2{\cdot}12 \times 10^{-3}}$$

$$= 4{\cdot}7 \times 10^{-12}\ \text{g ion l}^{-1}$$

$$\begin{aligned} pH &= -\log_{10} 4{\cdot}7 \times 10^{-12} \\ &= -[\log 4{\cdot}7 + \log 10^{-12}] \\ &= 11{\cdot}7 \end{aligned}$$

∴ The pH of 0·01 *M* methylamine is 11·7 (corr. to 3 sig. figs.).

(ii) Find K_b (NH_3) at 25°C if K_a (NH_4^+) is $6{\cdot}3 \times 10^{-10}$ mol l^{-1}.

$$\begin{aligned} K_b &= \frac{10^{-14}}{K_a} \\ &= \frac{10^{-14}}{6{\cdot}3 \times 10^{-10}} \\ &= 1{\cdot}6 \times 10^{-5} \text{ mol l}^{-1}. \end{aligned}$$

(iii) Find pK_b (HS^-) at 25°C if pK_a (H_2S) is 7·2.

$$pK_b = 14 - pK_a = 6{\cdot}8$$

(iv) Find the pH of a solution containing 80 cm³ of 0·1 *M* HCl mixed with 20 cm³ of 0·1 *M* NaOH. The solution contains the equivalent of 60 cm³ of 0·1 *M* HCl in 100 cm³ of solution or $\frac{60}{1000} \times 0{\cdot}1$ moles H^+ ions in 100 cm³ of solution.

$$\begin{aligned} \therefore [H^+(aq)] &= \frac{60 \times 0{\cdot}1}{1000} \times 10 \\ &= 6 \times 10^{-2} \text{ g ions l}^{-1}. \\ pH &= -\log_{10} 6 \times 10^{-2} \\ &= -[\log_{10} 6 + \log_{10} 10^{-2}] \\ &= -[0{\cdot}7782 - 2] \\ &= 1{\cdot}22 \end{aligned}$$

∴ The pH of the solution is 1·22 (corr. to 3 sig. figs.).

Measurement of pH

1 *Use of hydrogen electrodes*

The cell consists of a standard hydrogen electrode connected to a hydrogen electrode in the solution of unknown $[H^+(aq)]$. The e.m.f., *E*, of the cell is measured and the value of $[H^+(aq)]$ is found using the Nernst equation (Chapter 7, page 128).

$$E = E^{\ominus} + \frac{RT}{ZF} \log_e[H^+(aq)]$$

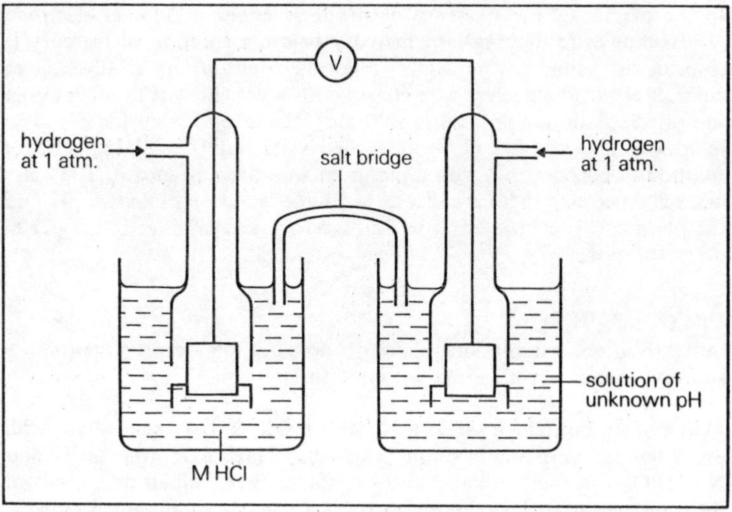

At 25°C this becomes

$$E = E^{\ominus} + 0{\cdot}059 \log_{10}[\mathrm{H^+(aq)}]$$

But for the standard hydrogen electrode $E^{\ominus}$ is 0.

$$\therefore E = -0{\cdot}069 \text{ pH}$$

However, hydrogen electrodes are inconvenient to use.

2 *pH meter*

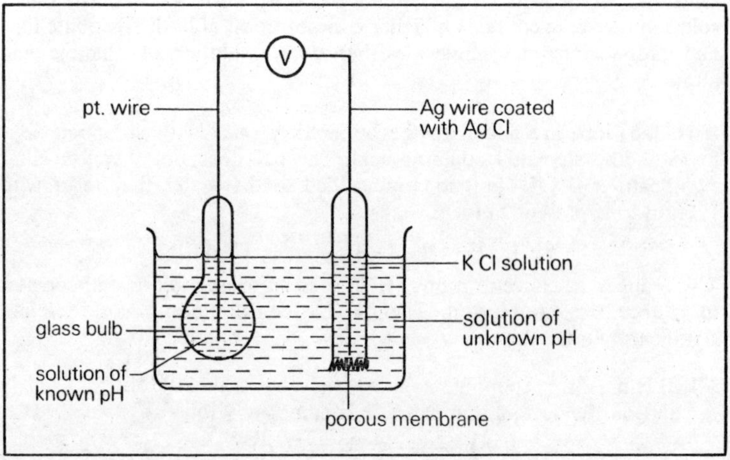

In the pH meter the reference electrode is either a calomel electrode (a platinum wire dipping into mercury below a solution of mercury(I) chloride in saturated potassium chloride solution) or a silver/silver chloride electrode (a silver wire coated with silver chloride dipping into a saturated solution of potassium chloride). The second electrode is a glass electrode. This consists of a thin walled glass bulb containing a buffer solution of constant pH with a platinum wire dipping into it. The glass and reference electrodes are placed in a buffer solution of known pH and the pH meter is calibrated. It is then used to read off directly the pH of other solutions.

Buffer solutions

Buffer solutions are solutions whose pH does not significantly alter on the addition of acids or alkalis or on dilution.

Acid buffers contain a weak acid and a salt of the same weak acid, e.g. ethanoic acid and sodium ethanoate, NaH_2PO_4 (the acid) and Na_2HPO_4 (the salt). Alkali buffers contain a weak alkali and a salt of the same weak base, e.g. ammonia solution and ammonium chloride.

Acid buffers, e.g. ethanoic acid and sodium ethanoate
In solution sodium ethanoate is completely ionized

$$CH_3COONa(aq) \longrightarrow CH_3COO^-(aq) + Na^+(aq).$$

The ethanoic acid is weak and so is only slightly ionized in solution. An equilibrium is set up.

$$CH_3COOH(aq) \rightleftharpoons CH_3COO^-(aq) + H^+(aq)$$

The large concentration of ethanoate ions from the salt depresses the ionization of the acid, i.e. the equilibrium is forced to the left. The buffer solution therefore contains a higher concentration of both ethanoate ions and undissociated acid molecules than does a solution of ethanoic acid alone.

If $H^+(aq)$ ions are added to the buffer they react with ethanoate ions to form undissociated ethanoic acid. The pH does not therefore alter significantly. If $OH^-(aq)$ ions are added to the buffer they react with $H^+(aq)$ ions present to form water.

$$H^+(aq) + OH^-(aq) \longrightarrow H_2O(l)$$

This reduces the concentration of $H^+(aq)$ so more ethanoic acid dissociates to restore the equilibrium. The pH of the solution is not therefore significantly altered.

Alkali buffers, e.g. ammonia solution and ammonium chloride
In solution the ammonium chloride is completely ionized

$$NH_4Cl(aq) \longrightarrow NH_4^+(aq) + Cl^-(aq).$$

The ammonia solution acts as a weak base so the following equilibrium is set up.

$$NH_3 \cdot H_2O(aq) \rightleftharpoons NH_4^+(aq) + OH^-(aq)$$

or $$NH_3(aq) + H^+(aq) \rightleftharpoons NH_4^+(aq)$$

The high concentration of $NH_4^+(aq)$ ions from the ammonium chloride forces the equilibrium to the left. The solution therefore contains a relatively high concentration of both ammonia molecules and ammonium ions.

If $H^+(aq)$ ions are added to the buffer solution, they react with ammonia molecules to form ammonium ions so that the pH does not alter significantly. If $OH^-(aq)$ ions are added to the buffer they react with $H^+(aq)$ ions present to form water.

$$H^+(aq) + OH^-(aq) \longrightarrow H_2O(l)$$

The $H^+(aq)$ ion concentration is restored by the equilibrium being forced to the left. The pH of the solution therefore does not significantly alter.

Calculations Consider a buffer solution containing ethanoic acid and sodium ethanoate. The following equilibrium is present.

$$CH_3COOH(aq) \rightleftharpoons CH_3COO^-(aq) + H^+(aq)$$

$$K_a = \frac{[CH_3COO^-(aq)][H^+(aq)]}{[CH_3COOH(aq)]}$$

or $$H^+(aq) = K_a \frac{[CH_3COOH(aq)]}{[CH_3COOH(aq)]}$$

$$\therefore -\log_{10} H^+(aq) = -\log_{10} K_a - \log_{10} \frac{[CH_3COOH(aq)]}{[CH_3COO^-(aq)]}$$

or $$pH = pK_a - \log_{10} \frac{[Acid]}{[Base]}$$

This equation also applies to alkaline buffers.

When [Acid] = [Base], pH = pK_a. Thus K_a of a weak acid may be found by measuring the pH of a solution of the acid which has been half neutralized by a strong base.

Examples

1 (a) Calculate the pH of a buffer solution which is 0·1 M with respect to both ethanoic acid and sodium ethanoate. K_a for ethanoic acid = $1{\cdot}8 \times 10^{-5}$ mol l^{-1}.

(b) Calculate the pH of 1 litre of the buffer solution in (a) after the addition of 10 cm^3 of 1 M HCl.

(a) Assuming that the ethanoic acid is undissociated and the salt is fully ionized, $[CH_3COOH(aq)] = 0{\cdot}1$ mol l^{-1} and

$$[CH_3COO^-(aq)] = 0{\cdot}1 \text{ g ion l}^{-1}.$$

$$\text{Using pH} = pK_a - \log_{10}\frac{[\text{Acid}]}{[\text{Base}]}$$

$$\therefore \text{pH} = -\log_{10}(1{\cdot}8 \times 10^{-5}) - 0$$

$$\therefore \text{pH} = 4{\cdot}74$$

(b) 10 cm³ of 1 *M* HCl are added, i.e. 0·01 g ions $H^+(aq)$ to 1 litre of the buffer. Therefore 0·01 moles of ethanoic acid are formed and 0·01 g ions of ethanoate ions are used up.

$$\therefore \text{New } [CH_3COOH(aq)] = 0{\cdot}1 + 0{\cdot}01$$

$$= 0{\cdot}11 \text{ mol l}^{-1}.$$

$$\text{New } [CH_3COO^-(aq)] = 0{\cdot}1 - 0{\cdot}01$$

$$= 0{\cdot}09 \text{ g ions l}^{-1}.$$

$$\therefore \text{New } [H^+] = \frac{1{\cdot}8 \times 10^{-5} \times 0{\cdot}11}{0{\cdot}09}$$

$$= 2{\cdot}2 \times 10^{-5} \text{ g ions l}^{-1}.$$

$$\therefore \text{New pH} = 4{\cdot}66.$$

2 In what proportions must 0·1 *M* solutions of ammonia and ammonium chloride be mixed to obtain a buffer solution of pH 9·8 (K_a NH_4^+ = 6×10^{-10} mol l^{-1}).

Using $$\text{pH} = pK_a - \log\frac{[\text{Acid}]}{[\text{Base}]}$$

$$\therefore 9{\cdot}8 = -\log(6 \times 10^{-10}) - \log\frac{[NH_4^+]}{[NH_3]}$$

$$\therefore 9{\cdot}8 = 9{\cdot}2 - \log\frac{[NH_4^+]}{[NH_3]}$$

$$\therefore \log\frac{[NH_4^+]}{[NH_3]} = -0{\cdot}6$$

$$= -1 + 0{\cdot}4$$

$$\therefore \frac{[NH_4^+]}{[NH_3]} = 0{\cdot}25$$

The 0·1 *M* ammonia and ammonium chloride solutions must be mixed in the ratio of 4 : 1 to produce a buffer of pH 9·8.

Use of buffer solutions

Many chemical processes, especially those occurring in living systems, are sensitive to changes in pH. Buffer solutions are therefore used to ensure that the pH remains fairly constant. For example, blood contains a number of buffering systems to maintain its pH at about 7·4. This prevents a decrease in pH on the dissolution of carbon dioxide.

Indicators

An acid–base indicator may itself be regarded as a weak acid for which either the acid and/or the base may be coloured. Letting HIn represent the indicator

$$\underset{\text{Colour A}}{\mathrm{HIn(aq)}} \rightleftharpoons \underset{\text{Colour B}}{\mathrm{H^+(aq) + In^-(aq)}}$$

On adding H^+(aq), the equilibrium is displaced to the left so colour A increases in intensity. On adding OH^-(aq), H^+(aq) are removed from solution as water is formed. The equilibrium is therefore displaced to the right and colour B increases in intensity.

For the indicator

$$\mathrm{pH} = pK_a - \log \frac{[\mathrm{HIn}]}{[\mathrm{In^-}]}$$

where K_a is the dissociation constant of the indicator.

When the indicator is on the point of changing colour [HIn] = [In^-]. The pH of the solution is then the pK_a of the indicator. Since different indicators have different equilibrium constants they change colour at different pH values as is shown in the following table. In practice a pH range of about 2 units is needed to change fully the colour of the indicator from its acidic to alkaline shade.

Indicator	K_a	Central pH	Alkaline colour	Acidic colour	Type of titration for which the indicator is suitable
Litmus	1×10^{-7}	7·0	Blue	Red	All acid–base titrations
Phenol-phthalein	2×10^{-10}	9·7	Magenta	Colour-less	Weak acid, strong base
Methyl orange	2×10^{-4}	3·7	Orange	Red	Strong acid, weak base

Titration curves

The change in pH during an acid-base titration may be followed using a pH meter.

(a) **Strong acid/strong base,** e.g. hydrochloric acid and sodium hydroxide

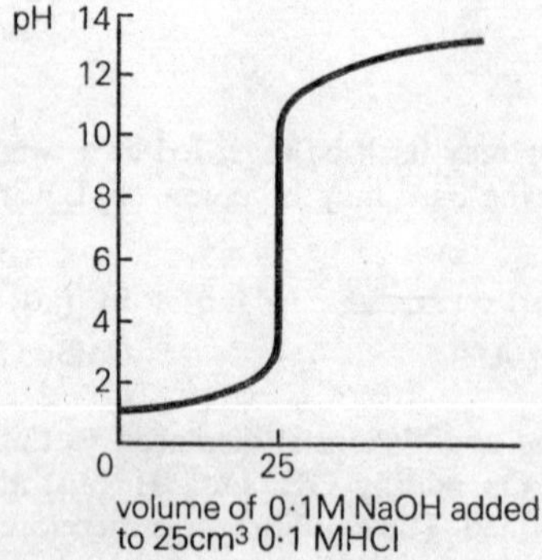

The pH changes slowly at first but very rapidly at the end point. Any indicator changing colour in the range pH 4 to pH 10 is suitable.

(b) **Weak acid/strong base,** e.g. ethanoic acid and sodium hydroxide

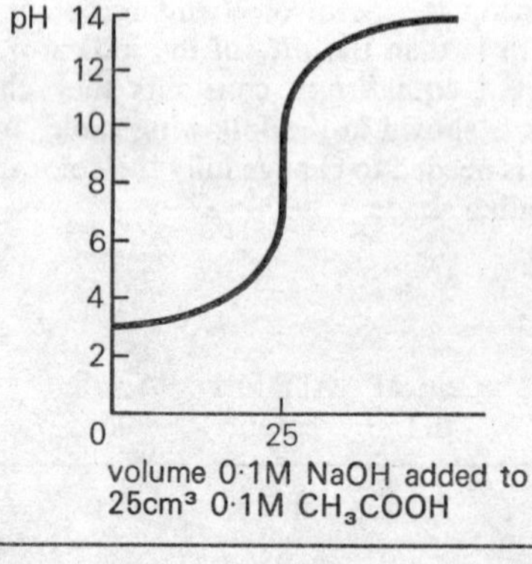

At the end point of the reaction the pH changes from about 7 to 11 and so phenolphthalein is the most suitable indicator. The salt formed, sodium ethanoate, is alkaline by hydrolysis.

(c) Strong acid/weak base, e.g. hydrochloric acid and ammonia solution

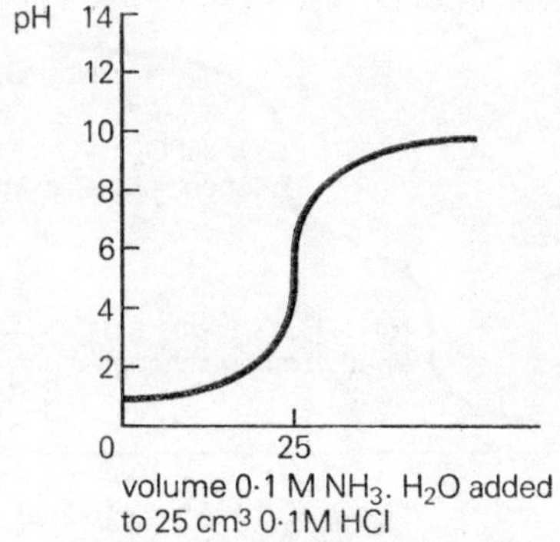

At the end point the pH changes from about 4 to 7 and so methyl orange is the most suitable indicator. The salt formed, ammonium chloride, is acidic by hydrolysis.

(d) Weak acid/weak base, e.g. ethanoic acid and ammonia solution

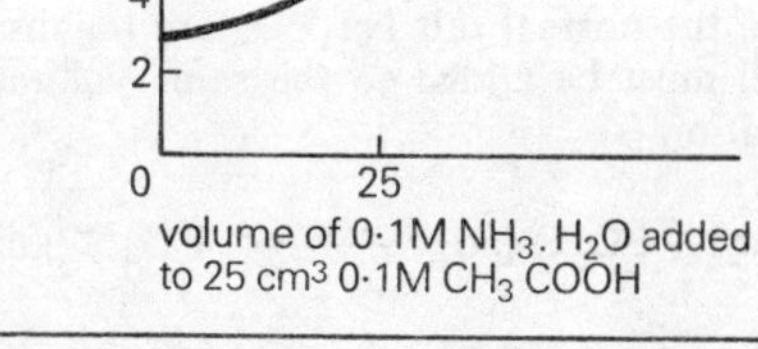

There is no sudden change in pH at the equivalence point and so no colour-change indicator is suitable. Estimation by titration is therefore not possible.

(e) Polybasic acid and strong base, e.g. phosphoric acid and sodium hydroxide

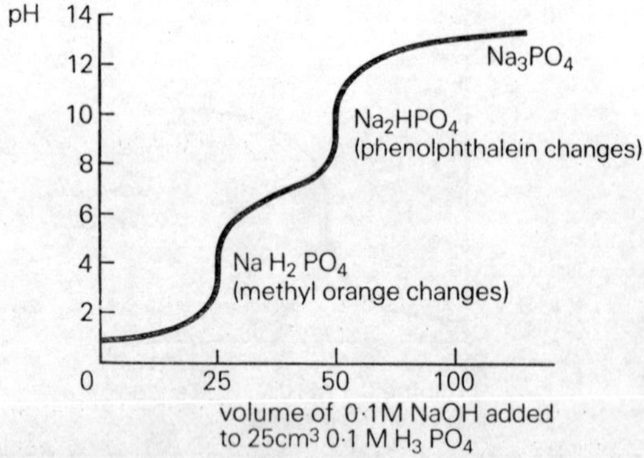

There are two rapid changes in pH but the third is not very distinct as the salt, Na_3PO_4, is completely hydrolysed.

(i) If the titration is carried out using methyl orange, then the solution changes colour when the following reaction is complete.

$$H_3PO_4(aq) + OH^-(aq) \longrightarrow H_2PO_4^-(aq) + H_2O(l)$$

Let the volume of alkali required be x cm^3.

(ii) If the titration is carried out using phenolphthalein, then the solution changes colour when the following reaction is complete.

$$H_3PO_4(aq) + 2OH^-(aq) \longrightarrow HPO_4^{2-}(aq) + H_2O(l)$$

The volume of alkali required is $2x$ cm^3.

(iii) If crystals of the normal salt Na_3PO_4 are required then $3x$ cm^3 of the alkali must be added to the same volume of acid as is used in titration (i).

$$H_3PO_4(aq) + 3OH^-(aq) \longrightarrow PO_4^{3-}(aq) + 3H_2O(l)$$

The solution is then boiled to remove water and left to crystallize.

(f) Polyacidic base and strong acid, e.g. sodium carbonate and hydrochloric acid

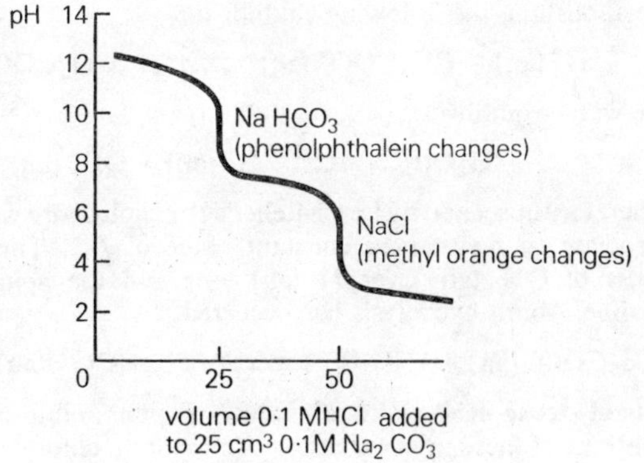

(i) If the titration is carried out using phenolphthalein as indicator then the solution changes from pink to colourless when the following reaction is complete.

$$Na_2CO_3 + HCl \longrightarrow NaHCO_3 + HCl$$

Let the volume of acid required be x cm^3.

(ii) If methyl orange is added to the colourless solution left in (i) and the titration is continued, then the solution changes from orange to red when the following reaction is complete.

$$NaHCO_3 + HCl \longrightarrow NaCl + CO_2 + H_2O$$

The volume of acid required is also x cm^3.

This method is used to find the amounts of sodium carbonate and sodium hydrogen carbonate in a mixture.

Hydrolysis of salts

(a) **Salts of strong acids and strong bases,** e.g. sodium chloride
Solutions of these salts are neutral as the equilibrium between the $H^+(aq)$ and $OH^-(aq)$ ions in the water is not upset. No hydrolysis therefore occurs.

$$NaCl(s) + aq \longrightarrow Na^+(aq) + Cl^-(aq)$$

$$H_2O(l) \rightleftharpoons H^+(aq) + OH^-(aq)$$

(b) **Salts of weak acids and strong bases,** e.g. sodium ethanoate, potassium carbonate, potassium cyanide, sodium sulphide
Solutions of these salts are alkaline, e.g. sodium ethanoate.
Sodium ethanoate is completely ionized in solution.

$$CH_3COONa(s) + aq \longrightarrow CH_3COO^-(aq) + Na^+(aq)$$

Ethanoic acid is a weak acid so protons are removed from solution in establishing the following equilibrium.

$$H^+(aq) + CH_3COO^-(aq) \rightleftharpoons CH_3COOH(aq)$$

The water equilibrium

$$H_2O(l) \rightleftharpoons H^+(aq) + OH^-(aq)$$

is therefore upset and by Le Chatelier's Principle more water molecules dissociate to maintain a constant value of K_w. This leads to an excess of $OH^-(aq)$ over $H^+(aq)$ ions and the solution becomes alkaline. Anion hydrolysis has occurred.

$$CH_3COO^-(aq) + H_2O(l) \rightleftharpoons CH_3COOH(aq) + OH^-(aq)$$

(c) **Salts of strong acids and weak bases,** e.g. ammonium chloride
Solutions of these salts are acidic. Ammonium chloride is completely ionized in solution.

$$NH_4Cl(s) + aq \longrightarrow NH_4^+(aq) + Cl^-(aq)$$

Ammonia solution is a weak base so the following equilibrium is set up.

$$NH_4^+(aq) + OH^-(aq) \rightleftharpoons NH_3 \cdot H_2O(aq)$$

The removal of $OH^-(aq)$ ions from solution upsets the water equilibrium.

$$H_2O(l) \rightleftharpoons H^+(aq) + OH^-(aq)$$

By Le Chatelier's Principle more water molecules dissociate to maintain the value of K_w so the solution becomes acidic.
Cation hydrolysis has occurred.

$$NH_4^+(aq) + H_2O(l) \rightleftharpoons NH_3 \cdot H_2O(aq) + H^+(aq)$$

(d) **Salts of weak acids and weak bases,** e.g. ammonium ethanoate
Solutions of these salts are extensively hydrolysed. The pH of the final solution depends on the relative values of K_a and K_b of the acid and base concerned, e.g. ammonium ethanoate which is completely ionized in solution.

$$CH_3COONH_4(s) + aq \longrightarrow CH_3COO^-(aq) + NH_4^+(aq)$$

The following equilibria are set up.

$$CH_3COO^-(aq) + H^+(aq) \rightleftharpoons CH_3COOH(aq)$$

$$NH_4^+(aq) + OH^-(aq) \rightleftharpoons NH_3 \cdot H_2O(aq)$$

Since pK_a (ethanoic acid) is 4·8 and pK_b (ammonia solution) is also 4·8, the solution is neutral. If pK_a of the acid is greater than pK_b of the base, then the solution is alkaline and vice versa.

(e) **Salts of aluminium and transition metals,** e.g. aluminium chloride, copper(II) sulphate, iron(III) chloride

Solutions of these salts are usually acidic as they are essentially salts of strong acids and weak bases. Equilibria of the following type are set up.

$$[Fe(H_2O)_6]^{3+}(aq) \rightleftharpoons [Fe(H_2O)_5OH]^{2+}(aq) + H^+(aq)$$

The position of the equilibrium depends on the charge to size ratio of the cation. The larger the value, the greater the polarizing effect of the ion on the co-ordinated water molecules and the more acidic the solution. Therefore solutions of Fe^{3+} salts are more acidic than those of Co^{2+} salts.

(f) **Acid salts,** e.g. sodium hydrogencarbonate, disodium hydrogenphosphate, sodium dihydrogenphosphate etc.
The pH of the solution of an acid salt depends on the extent of
(i) anion hydrolysis, and
(ii) dissociation of the hydrogen containing anion, e.g.
(1) Sodium hydrogencarbonate
(i) Anion hydrolysis

$$HCO_3^-(aq) + H_2O(l) \rightleftharpoons H_2CO_3(aq) + OH^-(aq)$$
$$pK_b = 7{\cdot}6$$

(ii) Dissociation of anion

$$HCO_3^-(aq) \rightleftharpoons H^+(aq) + CO_3^{2-}(aq)$$
$$pK_a = 10{\cdot}3$$

Since $pK_a > pK_b$, the solution is overall alkaline.
(2) Sodium hydrogensulphite
(i) Anion hydrolysis

$$HSO_3^-(aq) + H_2O(l) \rightleftharpoons H_2SO_3(aq) + OH^-(aq)$$

(ii) Dissociation of anion

$$HSO_3^-(aq) \rightleftharpoons SO_3^{2-}(aq) + H^+(aq)$$

Since $HSO_3^-(aq)$ ion is a stronger acid that it is a base the solution is acidic. This is also true for sodium hydrogensulphate.

(g) **Salts undergoing complete hydrolysis,** e.g. bismuth chloride, antimony chloride
When these salts are dissolved in hydrochloric acid and their solutions are diluted with water, a white precipitate of the oxysalt is obtained as complete hydrolysis occurs, e.g.

$$BiCl_3 + H_2O \rightleftharpoons BiOCl + 2HCl$$

The precipitate is redissolved by the addition of concentrated hydrochloric acid.

Practice questions

In questions 1 to 14 give the letter of the correct answer.

1 Which of the following describes the action of a Lewis acid? It
A. accepts a proton B. donates a proton
C. accepts an electron D. donates an electron
E. accepts an electron pair

2 Which of the following mixtures in aqueous solution produces a buffer solution?
A. NaOH and NaCl B. CH_3COONa and CH_3COOK
C. NaCl and HCl D. CH_3COOK and CH_3COOH
E. NH_4Cl and NH_4NO_3

3 K_a for methanoic acid at 298 K is 2×10^{-4} mol l^{-1}. The concentration of H^+ ions in g ion l^{-1} in 0·5 *M* methanoic acid solution at 298 K is
A. 1×10^{-4} B. 2×10^{-4} C. 4×10^{-4} D. 1×10^{-2}
E. 4×10^{-2}

4 Which of the following salts in aqueous solution produces an acidic solution?
A. Na_2S B. $AlCl_3$ C. $NaHCO_3$ D. Na_2CO_3
E. CH_3COOK

5 Two monoacidic bases have the following pK_b values at 25°C, base *X* $pK_b = 3{\cdot}0$, base *Y* $pK_b = 3{\cdot}8$. Which of the following is incorrect?
A. K_b for base *X* is 10^{-3}.
B. K_b for base *Y* is $1{\cdot}6 \times 10^{-4}$.
C. Base *Y* is stronger than base *X*.
D. The pK_b values change with temperature.
E. For solutions of equal concentration the pH of *X* is greater than that of *Y*.

6 In the equilibrium

$$HNO_3 + CH_3COOH \rightleftharpoons NO_3^- + (CH_3COOH_2)^+$$

which of the following pairs may be described as acids?
A. HNO_3 and $(CH_3COOH_2)^+$ B. CH_3COOH and NO_3^-
C. HNO_3 and NO_3^- D. CH_3COOH and $(CH_3COOH_2)^+$
E. HNO_3 and CH_3COOH

Questions 7 to 12

K_a for CH_3COOH and K_b for $NH_3 \cdot H_2O$ are both approximately 10^{-5} mol l^{-1} at 25°C. Choose from A to E the solution which best answers questions 7 to 12.

A. 1 *M* ethanoic acid
B. 1 *M* ammonia
C. A mixture of equal volumes of 1 *M* NH_4Cl + 1 *M* $NH_3 \cdot H_2O$
D. A mixture of equal volumes of 1 *M* CH_3COONa and 1 *M* CH_3COOH
E. A mixture of equal volumes of *M* CH_3COOH and *M* $NH_3 \cdot H_2O$

7 The solution with an approximate pH value of 7.

8 The solution with an approximate pH value of 5.

9 The most acidic solution.

10 The solution with pH nearest to 11·5.

11 The solution which forms an alkaline buffer.

12 K_a of CH_3COOH is $1{\cdot}8 \times 10^{-5}$ mol l^{-1} at 25°C. The H^+ ion concentration in g ion l^{-1} in 0·1 *M* CH_3COOH is
A. $\sqrt{1{\cdot}8 \times 10^{-5}}$ B. $\sqrt{1{\cdot}8 \times 10^{-6}}$ C. $\sqrt{1{\cdot}8 \times 10^{-4}}$
D. $1/\sqrt{1{\cdot}8 \times 10^{-6}}$ E. $\sqrt{1{\cdot}8} \times 10^{-6}$

13 Which of the following has the smallest pK_a value under a given set of conditions?
A. CH_2ICOOH B. $CH_2ClCOOH$ C. CH_3COOH
D. $C_6H_5CH_2COOH$ E. $C_2H_5CH_2COOH$

14 Equal volumes of the following aqueous solutions are mixed. Which solution will have the highest pH value?
A. 1 *M* NaOH 1 *M* CH_3COOH
B. 1 *M* KOH 1 *M* HCl
C. 1 *M* NaOH 1 *M* H_3PO_4
D. 1 *M* $NH_3{\cdot}H_2O$ 1 *M* H_2SO_4
E. 1 *M* $NH_3{\cdot}H_2O$ 0·5 *M* H_2SO_4

15 (a) Calculate the pH of the following solutions if $[H^+(aq)]$ is (i) 10^{-4} g ion l^{-1}, (ii) 0·03 g ion l^{-1}, (iii) $5{\cdot}3 \times 10^{-5}$ g ion l^{-1}.
(b) Calculate $[H^+(aq)]$ and $[OH^-(aq)]$ of the following solutions of pH values (i) 8, (ii) 0·2, (iii) 12·3.
(c) Calculate the pH of (i) 0·2 *M* NaOH, (ii) 0·02 *M* $Ba(OH)_2$, (iii) 0·5 *M* H_2SO_4, (iv) a mixture of 75 cm^3 0·1 *M* HCl and 25 cm^3 0·1 *M* NaOH, (v) 0·1 *M* CH_3COOH if $K_a = 1{\cdot}74 \times 10^{-5}$ mol l^{-1}, (vi) 0·001 *M* NH_2CH_2COOH if $K_a = 1{\cdot}7 \times 10^{-10}$ mol l^{-1}.

16 The dissociation of an indicator HIn in aqueous solution may be represented by

$$\text{HIn (blue)} \rightleftharpoons H^+ + In^- \text{ (colourless).}$$

(a) Write an expression for the dissociation constant K_a of the indicator.
(b) What is the colour of the indicator in aqueous sodium hydroxide solution?
(c) If the $[H^+]$ of the indicator is 4×10^{-3} in 0·1 *M* aqueous solution find K_a.
(d) Calculate pK_a of the indicator.

17

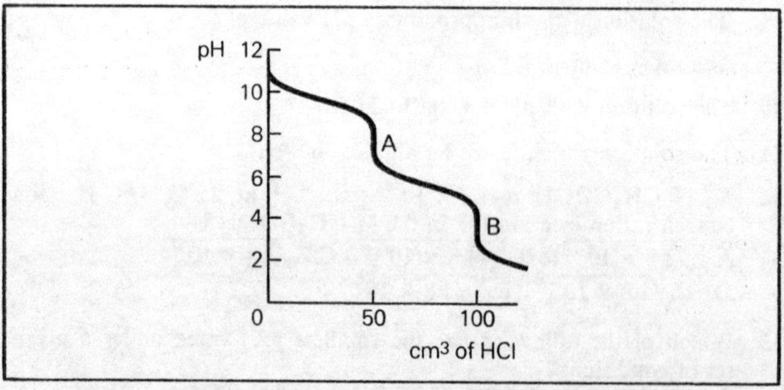

The graph shows the plot of pH against the volume of 0·04 *M* HCl added to 40 cm^3 of a solution of an alkali metal carbonate.

(a) Give the reaction completed at *A* and an indicator for this point.

(b) Give the reaction completed at *B* and an indicator for this point.

(c) Comment on the rate of change in pH in the region near *A*.

(d) What is the molarity of the carbonate solution?

(e) What difficulty would arise if the titration was carried out using 0·04 *M* CH_3COOH instead of 0·04 *M* HCl?

(f) Explain why the $[Al(H_2O)_6]^{3+}$ ion is a stronger acid than the $[Mg(H_2O)_6]^{2+}$ ion.

9 Thermodynamics

Thermodynamic arguments may be used to predict whether or not a chemical change will tend to take place under stated conditions. It does not give information as to the rate of the reaction.

First law of thermodynamics (see chapter 3).

Standard free energy change $\Delta G^{\ominus}_{298}$

Both exothermic and endothermic reactions may occur spontaneously, e.g.

(a) The exothermic combination of hydrogen and oxygen occurs spontaneously in the presence of platinum

$$2H_2(g) + O_2(g) \longrightarrow 2H_2O(l)$$

$$\Delta H^{\ominus} = -572 \text{ kJ mol}^{-1}$$

(b) The endothermic reaction between sulphur dichloride oxide and iron(III) chloride-6-water occurs spontaneously when the reagents are mixed.

$$6SOCl_2(l) + FeCl_3.6H_2O(s) \longrightarrow FeCl_3(s) + 6SO_2(g) + 12HCl(g)$$

$$\Delta H^{\ominus} = +1271 \text{ kJ mol}^{-1}$$

Therefore the sign of the enthalpy change is not a reliable indicator of the direction of a spontaneous chemical reaction.

However, in general, for any chemical reaction to take place spontaneously there must be a loss in energy called the free energy ΔG. The value of ΔG depends on temperature and pressure and so the standard free energy change $\Delta G^{\ominus}_{298}$ is defined. This is the value of ΔG at 298 K and 1 atmosphere pressure when any elements concerned are in their normal state under those conditions.

For a chemical reaction to occur spontaneously $\Delta G^{\ominus}_{298}$ must be negative.

$\Delta G^{\ominus}_{298}$ is a measure of the maximum work obtainable from a reaction carried out at constant pressure (ignoring work done on expansion).

For any reaction

$$\Delta G^{\ominus}_{298} = \sum \Delta G^{\ominus}_{298} \text{ (products)} - \sum \Delta G^{\ominus}_{298} \text{ (reactants)}$$

If (a) $\Delta G^{\ominus}_{298} = 0$, then the system is in equilibrium.

(b) $\Delta G^{\ominus}_{298}$ is negative, then the reaction occurs spontaneously.

(c) $\Delta G^{\ominus}_{298}$ is positive, then the reaction cannot occur under the standard conditions.

The standard free energy of formation of a compound X, $\Delta G^{\ominus}_{f,298}(X)$, is the standard free energy change for the reaction in which 1 mole of the compound is formed from its elements in their normal states at 298 K and 1 atmosphere pressure. $\Delta G^{\ominus}_{f,298}$ of an element is zero.

Example
For the reaction

$$C(s) + O_2(g) \longrightarrow CO_2(g) \quad \Delta G^{\ominus}_{298} = -395 \text{ kJ}$$

$$\Delta G^{\ominus}_{298} = \sum \Delta G^{\ominus}_{298}(\text{products}) - \sum \Delta G^{\ominus}_{298}(\text{reactants})$$

$$\therefore -395 = \Delta G^{\ominus}_{f,298}(CO_2(g)) - [\Delta G^{\ominus}_{f,298}(O_2(g)) + \Delta G^{\ominus}_{f,298}(C(s))]$$

$$= \Delta G^{\ominus}_{f,298}(CO_2(g)) - 0$$

$$\therefore \Delta G^{\ominus}_{f,298}(CO_2(g)) = -395 \text{ kJ}$$

$\Delta G^{\ominus}_{298}$ for redox reactions
The energy change $\Delta G^{\ominus}_{298}$ during a redox reaction is related to the e.m.f. of the cell $E^{\ominus}$ by

$$\Delta G^{\ominus}_{298} = -zFE^{\ominus}$$

where z is the number of electrons transferred and F is the Faraday.

Example
For the cell

$$Cu(s)|Cu^{2+}(aq)\,\vdots\,Ag^{+}(aq)|Ag(s) \qquad E^{\ominus} = +0{\cdot}46 \text{ V}$$

The reaction occurring is

$$Cu(s) + 2Ag^{+}(aq) \longrightarrow 2Ag(s) + Cu^{2+}(aq)$$

Using

$$\Delta G^{\ominus}_{298} = -zFE^{\ominus}$$

$$\Delta G^{\ominus}_{298} = -2 \times 96500 \times 0{\cdot}46$$

$$= -88 \text{ kJ mol}^{-1}$$

For redox reactions the sign of the $E^{\ominus}$ value for the particular system may be used to predict the direction of a chemical change (see chapter 7).

$\Delta G^{\ominus}$ for equilibrium reactions
For equilibrium reactions, $\Delta G^{\ominus}$ is related to the equilibrium constant K by the relationship

$$\Delta G^{\ominus} = -RT \log_e K$$

or

$$\Delta G^{\ominus} = -2{\cdot}3RT \log_{10} K$$

where R is the gas constant and T is the temperature in K.

Example

Calculate the equilibrium constant for the equilibrium

$$N_2(g) + 3H_2(g) \rightleftharpoons 2NH_3(g)$$

given $\Delta G^{\ominus}_{f,\,298}(NH_3(g)) = -17\ \text{kJ mol}^{-1}$,

$$\Delta G^{\ominus}_{298} = 2\Delta G^{\ominus}_{f,\,298}(NH_3(g)) - \Delta G^{\ominus}_{f,\,298}(N_2(g)) - 3\Delta G^{\ominus}_{f,\,298}(H_2(g)).$$

The standard free energies of formation of elements for their standard states are 0.

$$\therefore \Delta G^{\ominus}_{298} = 2 \times (-17)$$
$$= -34\ \text{kJ}$$

but
$$\Delta G^{\ominus} = -2{\cdot}3RT \log_{10} K_p$$
$$\therefore \Delta G^{\ominus}_{298} = -2{\cdot}3 \times 8{\cdot}3 \times 10^{-3} \times 298 \log K_p$$
$$\therefore -34 = -5{\cdot}69 \log K_p$$
$$\therefore \log K_p = 5{\cdot}98$$
$$K_p = 9{\cdot}6 \times 10^5\ \text{atm}^{-2} \text{ at } 298\ \text{K}.$$

Thus, if a catalyst could be found and equilibrium could be reached, then equilibrium mixtures of hydrogen and nitrogen at 298 K and 1 atmosphere pressure would be expected to contain mainly ammonia and only small amounts of nitrogen and hydrogen.

Entropy

For any spontaneous reaction, $\Delta G^{\ominus}$ and $\Delta H^{\ominus}$ are related by the equation

$$\Delta G^{\ominus} = \Delta H^{\ominus} - T\Delta S^{\ominus}$$

where T is the temperature in K and $\Delta S^{\ominus}$ is the change in entropy. Entropy may be regarded as a measure of the extent to which a system is disordered. It usually increases with temperature as the average kinetic energy of the particles in the system increases with temperature. Thus when a crystalline solid melts there is an increase in entropy. An increase in entropy is an increase in the number of ways in which the energy of the system can be distributed amongst its energy levels.

If a spontaneous reaction is endothermic, then for $\Delta G^{\ominus}$ to be negative there must be a large increase in entropy, e.g. the dissolution of ammonium nitrate in water is endothermic. During the process there is a large increase in entropy as the particles in the solution are more disordered than in the crystal.

Measurement of entropy change

For a reversible (isothermal) process carried out at a fixed temperature, the change in entropy is equal to the heat absorbed or evolved at constant pressure divided by the temperature, i.e.

$$\Delta S = \frac{\Delta H}{T}$$

If heat is absorbed, ΔS is positive, i.e. there is an increase in entropy.

If heat is evolved, ΔS is negative, i.e. there is a decrease in entropy.

Example

Find the entropy change when ice melts if the molar latent heat of fusion of ice is 6·025 kJ mol^{-1}.

Ice melts at 273 K

$$\therefore \Delta S = \frac{6{\cdot}025}{273}$$

$$= 0{\cdot}02207 \text{ kJ K}^{-1} \text{ mol}^{-1}.$$

$\therefore$ When ice melts there is an increase in entropy of 22·07 J K^{-1} mol^{-1}.

Equilibrium reactions

1 $$CaCO_3(s) \rightleftharpoons CaO(s) + CO_2(g)$$

$$\Delta H^\ominus = +178 \text{ kJ}$$

Since $\Delta G^\ominus = \Delta H^\ominus - T\Delta S^\ominus$, $T\Delta S^\ominus$ must be sufficiently large to counteract the positive value of $\Delta H^\ominus$ in order to make $\Delta G^\ominus$ negative. Dissociation will then occur. Since $\Delta S^\ominus$ is positive, $\Delta G^\ominus$ will only be negative at high temperatures. In practice calcium carbonate dissociates only at high temperatures.

2 $$N_2O_4(g) \rightleftharpoons 2NO_2(g)$$

$$\Delta H^\ominus = +58{\cdot}2 \text{ kJ}$$

$\Delta S^\ominus$ for the forward reaction is positive but since $\Delta H^\ominus$ is relatively small only a small value of T is needed to make $\Delta G^\ominus$ negative. In practice the equilibrium is well over to the right at low temperatures.

3 $$N_2(g) + 3H_2(g) \rightleftharpoons 2NH_3(g)$$

$$\Delta H^\ominus = -92 \text{ kJ}$$

$\Delta S^\ominus$ for the forward reaction is negative so for $\Delta G^\ominus$ to be negative only small values of T are needed. In practice the equilibrium is well over to the right at low temperatures.

In general, if there is an increase in gaseous volume on reaction, ΔS is positive and if there is a decrease ΔS is negative.

Practice questions

1 Predict whether the following changes involve an increase or decrease in entropy.

a. $Cl_2(g) \longrightarrow Cl_2(l)$
b. $Br_2(g) \longrightarrow 2Br(g)$
c. $2CO(g) + O_2(g) \longrightarrow 2CO_2(g)$
d. the crystallization of salt from brine
e. $C(s) + ZnO(s) \longrightarrow CO(g) + Zn(s)$
f. $2Al(s) + 3Fe^{2+}(aq) \longrightarrow 2Al^{3+}(aq) + 3Fe(s)$

2 In the reaction

$$Cu(s) + 2Ag^+(aq) \rightleftharpoons Cu^{2+}(aq) + 2Ag(s)$$

the following values are obtained at 25°C. $\Delta H = -121$ kJ, $E^\ominus = 0{\cdot}46$ V. Calculate (i) $\Delta G^\ominus$, (ii) $\Delta S^\ominus$.

3 In the reaction

$$CaCO_3(s) \rightleftharpoons CaO(s) + CO_2(g)$$

calculate the equilibrium pressure of CO_2 over $CaCO_3$ at 25°C given the $\Delta G^\ominus_{f,\,298}$ values of $CO_2(g)$, $CaO(s)$ and $CaCO_3(s)$ are -395, -604 and -1130 kJ mol^{-1} respectively.

4 Given the standard free energies of formation of the following ions

Ions	$Fe^{3+}(aq)$	$Fe^{2+}(aq)$	$Ag^+(aq)$
$\Delta G^\ominus_{f,\,298}$ in kJ mol^{-1}	$-10{\cdot}7$	$-84{\cdot}9$	$+77{\cdot}2$

Find the equilibrium constant at 25°C for the reaction

$$Ag^+(aq) + Fe^{2+}(aq) \rightleftharpoons Fe^{3+}(aq) + Ag(s).$$

5 The maximum e.m.f. of a Daniell cell in which the reaction

$$Zn(s) + Cu^{2+}(aq) \longrightarrow Zn^{2+}(aq) + Cu(s)$$

$$\Delta H = -210 \text{ kJ}$$

takes place is 1·107 V. Calculate the entropy change if all values given are at 25°C.

6 Given the following:

	$N_2O(g)$	$NH_3(g)$	$H_2O(g)$
$\Delta G^\ominus_{f,\,298}$ kJ mol^{-1}	104	$-16{\cdot}5$	-228

Use the data to calculate (i) $\Delta G^\ominus$ and (ii) K_p for the following reaction at 298 K.

$$N_2O(g) + 4H_2(g) \rightleftharpoons 2NH_3(g) + H_2O(g)$$

7 For the reaction

$$H_2O(g) + C(s) \rightleftharpoons H_2(g) + CO(g)$$

at 1300 K, $\Delta H^{\ominus}_{1300} = +146{\cdot}9$ kJ and $\Delta G^{\ominus}_{1300} = -57{\cdot}6$ kJ.

(a) Calculate (i) log K_p, (ii) K_p at 1300 K.

(b) Calculate K_p at 1500 K.

(c) At which of these temperatures is the equilibrium yield of $H_2(g)$ and $CO(g)$ greater?

10 Chemical Kinetics

Chemical kinetics is concerned with the study of rates of chemical reactions and can provide information about their mechanisms.

Rate of reaction

The rate of chemical reaction is investigated by measuring the decrease in concentration of reactants or the increase in concentration of products with time. Experimentally this may involve measuring with time

(i) the volume of a gas evolved,

(ii) a colour change using a colorimeter,

(iii) a change in electrical resistance, refractive index, optical rotation or volume of a solution.

An alternative approach involves the withdrawal of samples from the reaction mixture at regular intervals. The samples are immediately quenched by dilution, cooling etc. to stop the reaction and the concentration of the reactants or products in the sample is found by titration. The rate of hydrolysis of an ester such as methyl ethanoate by hydroxide ions may be investigated as follows.

$$CH_3COOH_3(aq) + OH^-(aq) \longrightarrow CH_3COO^-(aq) + CH_3OH$$

Small measured portions of the reaction mixture are pipetted out at regular intervals and run into a measured excess of dilute acid to stop the reaction. The excess acid is titrated with standard alkali to find the amount of hydroxide ions used up in the hydrolysis.

Measurements of this type do not give the rate of reaction directly but only the concentration of reactants or products at a particular time. A graph of concentration against time is plotted and the rate of a reaction at a certain time is given by the gradient of the tangent at that point.

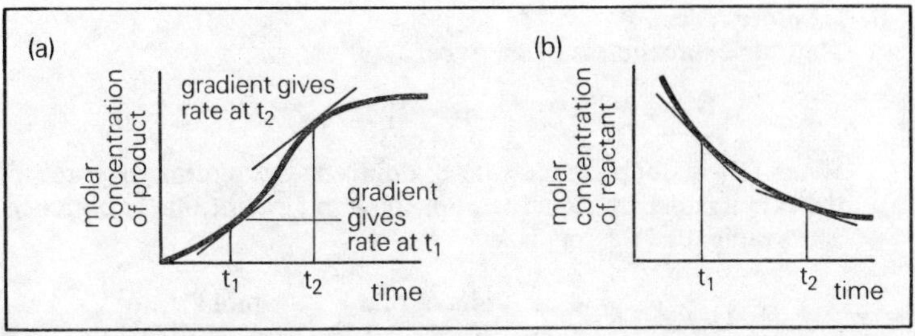

From this data further graphs are plotted of, for example, the rate against the concentration of a certain reactant to find the order of the reaction. The rate of the reaction is affected by

(a) the concentration of reactants,
(b) the temperature,
(c) the physical nature of the reactants,
(d) the presence of catalysts,
(e) the wavelength and intensity of incident radiation (for certain reactions only).

(a) Concentration
The overall rate of a chemical reaction is in general increased by an increase in concentration of one or more of the reactants. The mathematical relationship between rate and concentration is found by experiment. For a reaction with an equation of the type

$$mA + nB \longrightarrow \text{products}$$

the initial rate may be found by experiment to be proportional to $[A]^x$ and $[B]^y$. Then the rate expression for the reaction is given by the relationship

$$\text{rate} = k[A]^x[B]^y$$

where k is known as the rate constant (it is constant for a given reaction at a given temperature) and x and y may be whole numbers or fractions.

The reaction is said to be x order with respect to A, y order with respect to B and $(x + y)$ order overall. The overall order of a reaction is the sum of the indices of the concentration terms in the experimentally determined rate expression.

Note: The rate expression cannot be determined from the stoichiometric equation.

(i) 1st order reactions
The rate expression is of the type

$$\text{rate} = k[A].$$

When $[A]$ is doubled the rate is doubled. On plotting a graph of the rate against the concentration of A, a straight line is obtained. The gradient of the line is k.

$$\text{Units of } k = \frac{\text{units of rate}}{\text{units of concentration}} = \frac{\text{mol l}^{-1}\text{ s}^{-1}}{\text{mol l}^{-1}}.$$

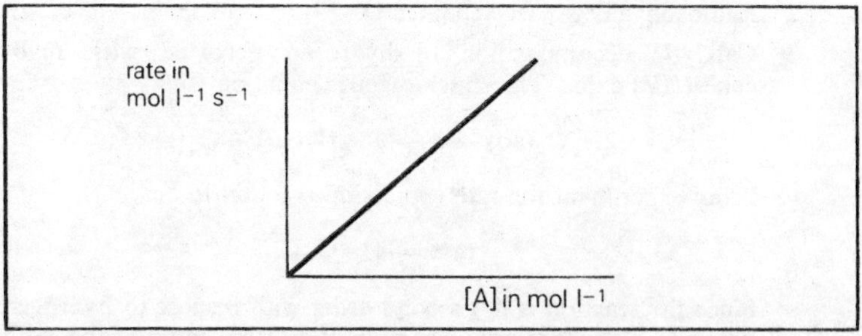

A first order reaction has a half-life which is independent of the initial concentration of reactants. The half-life, $t_{\frac{1}{2}}$ is the time taken for half of the original amount of reactants to be used up. On plotting a graph of concentration against time the curve shown is obtained.

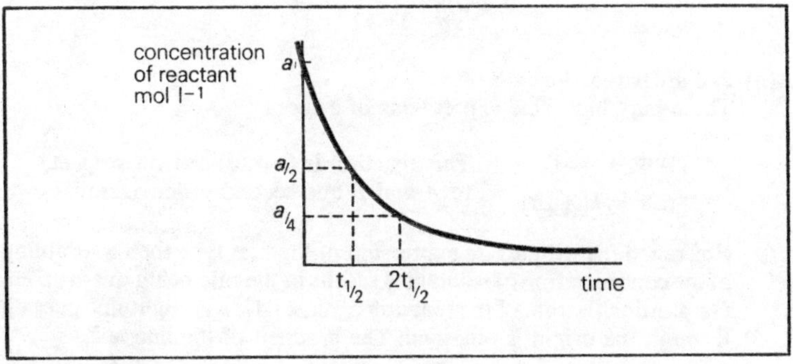

Examples

1 The inversion of cane sugar

$$\underset{\substack{\text{sucrose} \\ \text{(dextro)}}}{C_{12}H_{22}O_{11}} + H_2O \longrightarrow \underbrace{\underset{\substack{\text{fructose} \\ \text{(laevo)}}}{C_6H_{12}O_6} + \underset{\substack{\text{glucose} \\ \text{(dextro)}}}{C_6H_{12}O_6}}_{\text{invert sugar (laevo)}}$$

The optical activity of fructose is greater than that of glucose so the mixture is overall laevorotatory. Therefore, as the reaction proceeds the polarized light is rotated less and less to the right and eventually is rotated to the left. The amount of rotation is measured at specific times using a polarimeter and gives a measure of the amount of sugar used up. From the results of the experiment it is found that

$$\text{rate of reaction} = k\,[\text{sucrose}].$$

2 Radioactive decay (see chapter 1).

3 Catalytic decomposition of hydrogen peroxide with manganese(IV) oxide. The stoichiometric equation is

$$2H_2O_2(aq) \longrightarrow 2H_2O(l) + O_2(g).$$

From experiment the rate expression is found to be

$$\text{rate} = k[H_2O_2].$$

Since the reaction is not second order with respect to hydrogen peroxide it cannot involve straightforward collision between 2 molecules of hydrogen peroxide with subsequent decomposition to products. A possible mechanism could involve 2 steps as indicated, the first being slow and rate determining.

$$H_2O_2(aq) \longrightarrow H_2O(l) + O(g)$$
$$2O(g) \longrightarrow O_2(g)$$

(ii) 2nd order reactions

These may have rate expressions of 2 types

$\text{rate} = k[A]^2$

$\text{rate} = k[A][B]$ — This reaction is first order with respect to A and B and second order overall.

For reactions with a rate expression of the first type then a doubling of the concentration of reactant A results in the rate being quadrupled. On plotting the rate of the reaction against $[A]^2$ a straight line passing through the origin is obtained. The gradient of the line is k.

$$\text{Units of } k = \frac{\text{mol l}^{-1}\text{ s}^{-1}}{\text{mol}^2\text{ l}^{-2}} = \text{l mol}^{-1}\text{ s}^{-1}.$$

The half-life of a second order reaction depends on the initial concentration.

Examples

1 The vapour phase decomposition of ethanal

$$CH_3CHO(g) \longrightarrow CH_4(g) + CO(g)$$
$$\text{rate} = k[CH_3CHO(g)]^2$$

2 Oxidation of iodine ions by hydrogen peroxide in acid solution.

$$H_2O_2(aq) + 2H^+(aq) + 2I^-(aq) \longrightarrow 2H_2O(l) + I_2(aq)$$
$$\text{rate} = k[H_2O_2(aq)][I^-(aq)]$$

It is suggested that the mechanism may involve a slow, rate-determining step of the type

$$H_2O_2(aq) + I^-(aq) \longrightarrow H_2O(l) + OI^-(aq)$$

followed by fast reactions of the type

$$H^+ + OI^- \longrightarrow HOI$$

$$HOI + H^+ + I^- \longrightarrow I_2 + H_2O$$

3 Alkaline hydrolysis of ethyl ethanoate

$$CH_3COOC_2H_5(aq) + OH^-(aq) \longrightarrow C_2H_5OH(aq) + CH_3COO^-(aq)$$

$$\text{rate} = k[CH_3COOC_2H_5(aq)][OH^-(aq)]$$

(iii) 3rd order reactions

These may have rate expressions of various types, e.g.

$$\text{rate} = k[A]^3$$

$$\text{rate} = k[A]^2[B]$$

$$\text{rate} = k[A][B][C]$$

For reactions with a rate expression of the first type, a doubling in the concentration of A results in an 8-fold increase in the rate. A graph of rate against $[A]^3$ produces a straight line passing through the origin of gradient k.

$$\text{Units of } k = \frac{\text{mol l}^{-1}\text{ s}^{-1}}{\text{mol}^3\text{ l}^{-3}} = \text{l}^2\text{ mol}^{-2}\text{ s}^{-1}.$$

3rd order reactions are rare.

(iv) Reactions with a fractional order

Some reactions usually involving more than one step may have a fractional order, e.g.

$$CO(g) + Cl_2(g) \longrightarrow COCl_2(g)$$

$$\text{rate} = k[CO][Cl_2]^{\frac{1}{2}}$$

Molecularity

The molecularity of a reaction is the number of molecules (or ions) which react together in the rate determining step. Most reactions are bimolecular, fractional molecularities are impossible.

The molecularity of a reaction cannot be determined from the stoichiometric equation. Consider the following data.

Initial concentration of reactant X in mol l^{-1}	Initial concentration of reactant Y in mol l^{-1}	Initial reaction rate in mol l^{-1} h^{-1}
1	0·3	$3{\cdot}4 \times 10^{-10}$
1	0·6	$1{\cdot}4 \times 10^{-9}$
1	0·9	$3{\cdot}1 \times 10^{-9}$
2	0·9	$6{\cdot}2 \times 10^{-9}$
3	0·9	$9{\cdot}3 \times 10^{-9}$

(a) Find the order of the reaction with respect to (i) X and (ii) Y.

(b) Give the experimental rate law and calculate the value of the rate constant.

(c) Stoichiometric experiments indicate that the balanced equation for the reaction is

$$2X(g) + 2Y(g) \longrightarrow 2Z(g) + W(g).$$

Suggest a possible mechanism.

(a) (i) The reaction is 1st order with respect to X as when the concentration of X is doubled with that of Y held constant, the rate is doubled.

(ii) The reaction is 2nd order with respect to Y as when the concentration of Y is doubled with that of X held constant, the rate is quadrupled.

(b) The experimental rate law is

$$\text{rate} = k[X][Y]^2$$

$$\therefore k = \frac{\text{rate}}{[X][Y]^2}$$

Using the first set of results

$$k = \frac{3{\cdot}4 \times 10^{-10}}{1 \times (0{\cdot}3)^2} = 3{\cdot}77 \times 10^{-9}\,\frac{\text{mol}\,l^{-1}\,h^{-1}}{\text{mol}^3\,l^{-3}}$$

$$\therefore k = 3{\cdot}77 \times 10^{-9}\ l^2\ \text{mol}^{-2}\ h^{-1}$$

(c) A possible mechanism could be

$$2Y \longrightarrow A \quad \text{slow}$$
$$A + X \longrightarrow B \quad \text{slow}$$
(the two slow steps above: rate determining steps)
$$B + X \longrightarrow 2Z + W \quad \text{fast}$$

(b) Temperature

The rate of a reaction approximately doubles for every 10°C rise in temperature. This observation may be explained by applying the **collision theory** of reaction kinetics.

Collision theory

Many gaseous bimolecular reactions of the type

$$A(g) + B(g) \longrightarrow \text{products}$$

have similar rate constants at a given temperature. This indicates that the rate of these reactions must be related to the rate of collision between the reacting species. However, kinetic experiments show that not all collisions result in product formation, i.e. not all collisions are **effective**. For example, at 750 K, the number of effective collisions per second (found from rate measurements) is only about 10^{-11}% of all collisions per second (found from the kinetic theory). Since a rise in temperature of 10 K approximately doubles the rate of reaction, the rate of effective collisions must be doubled. However, the total number of collisions is only slightly increased by that temperature rise.

These observations are explained by assuming that the reaction will only occur if molecules with an energy greater than a certain minimum value collide together. This minimum energy is called the activation energy E_A and is the energy required for bonds to begin to break before others can be made. The energy changes during the course of an endothermic and exothermic reaction are shown below.

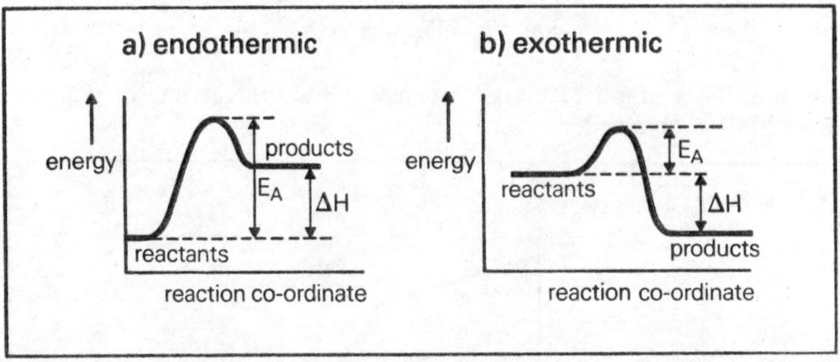

The graph below illustrates the distribution of molecular energies within a given sample of gas at temperatures T_1 K and $(T_1 + 10)$ K.

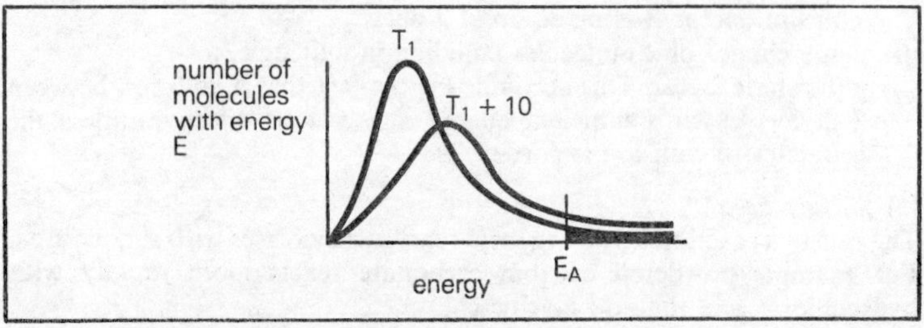

At T_1 K, the number of molecules with an energy of at least E_A is proportional to the shaded area. This area approximately doubles when the temperature is raised by 10°C.

From Boltzmann's law

$$\text{No. of molecules with an energy } E \text{ or greater} = n_0 e^{-E/RT}$$

where R is the gas constant, T is the temperature in K and n_0 is the total number of molecules.

$$\therefore \text{ The fraction of molecules with energy } E \text{ or greater} = e^{-E/RT}$$

Arrhenius proposed that the reaction rate measured in terms of the rate constant is proportional to the fraction of molecules having an energy of E_A or greater, i.e.

$$k = Ae^{-E_A/RT}$$

where A is known as the collision factor.

$$\therefore \log_e k = \log_e A - \frac{E_A}{RT}$$

or

$$2{\cdot}303 \log_{10} k = 2{\cdot}303 \log_{10} A - \frac{E_A}{RT}$$

or

$$\log_{10} k = \log_{10} A - \frac{E_A}{2{\cdot}303RT}$$

On plotting a graph of $\log_{10} k$ against $1/T$, a straight line of gradient $E_A/2{\cdot}303R$ is obtained.

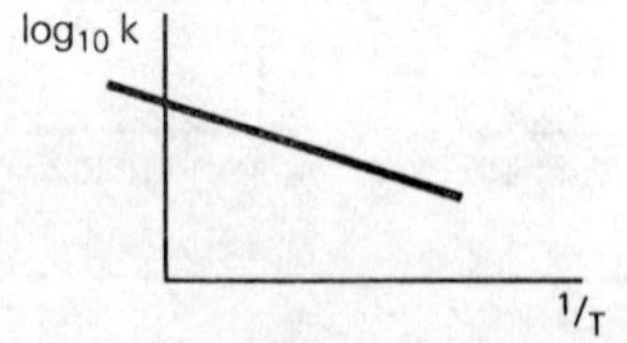

E_A for the reaction is therefore found.
The collision factor A is made up of 2 parts

(i) Z the chance of 2 molecules colliding in unit time.
(ii) p the steric factor. This accounts for the fact that a collision between 2 molecules with sufficient energy may not be effective unless the geometry of collision is correct.

(c) *Surface area*
The rate of a reaction involving solid reactants increases with surface area. For example powdered calcium carbonate reacts more rapidly with hydrochloric acid than do marble chips.

(d) Catalysts
A catalyst is a substance which increases the rate of a reaction without itself undergoing any quantitative change. However, the catalyst may be altered physically during the course of a reaction, e.g. granular manganese(IV) oxide becomes finely powdered when it is used as a catalyst in the decomposition of potassium chlorate. The rate constant k of the reaction is altered.

Homogeneous catalysis – the catalyst and reactants are in the same phase, e.g. the use of H^+(aq) ions to catalyse esterification reactions of the type

$$C_2H_5OH(aq) + CH_3COOH(aq) \xrightleftharpoons{H^+(aq)} CH_3COOC_2H_5(aq) + H_2O(l).$$

Heterogeneous catalysis – the catalyst and reactants are in different phases, e.g. the use of nickel in the gaseous hydrogenation of alkenes.

$$H_2(g) + C_2H_4(g) \xrightarrow{Ni} C_2H_6(g)$$

Note: A catalyst does not alter the value of K or the position of equilibrium in equilibrium reactions. It increases the rate of both forward and backward reactions so that equilibrium is reached more quickly. A catalyst lowers the activation energy of a reaction by providing one or more alternative pathways. The effect on the energy profile is shown below.

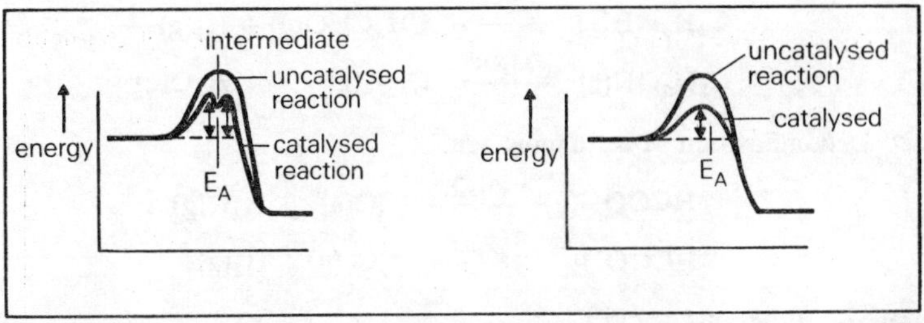

Theories of catalysis

1 **Intermediate compound theory** (energy profile (i))
During the reaction the catalyst forms an unstable compound or intermediate which then decomposes. For example it has been suggested that during the contact process, the catalyst V_2O_5 forms an unstable intermediate of the formula V_2O_4.

$$SO_2(g) + V_2O_5(s) \longrightarrow SO_3(g) + V_2O_4(s)$$
$$O_2(g) + 2V_2O_4(s) \longrightarrow 2V_2O_5(s)$$

Many catalysts forming intermediate compounds are compounds of transition metals as the metal can exert a variable valency.

2 Adsorption theory (energy profile (ii))
In this theory it is assumed that one or more reactants are adsorbed onto the surface of the catalyst with the formation of weak bonds. After reaction these bonds are broken.

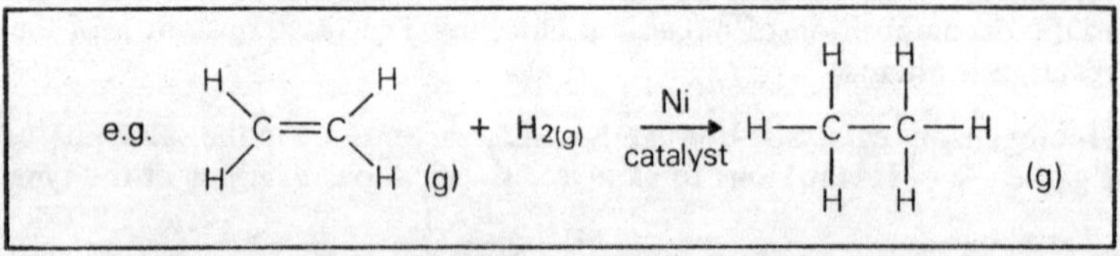

The hydrogen molecules are adsorbed on the surface of the nickel catalyst and as the bonds between the metal and hydrogen atoms strengthen, the H—H bonds in the hydrogen molecules weaken. Bonds form between the ethene molecule and adsorbed hydrogen atoms. The bonds between the metal and hydrogen atoms are broken as the new C—H bonds are formed.

Effect of catalysts on the route of a reaction
A particular catalyst may catalyse only 1 of the alternative routes available to a certain reactant, e.g.

1 Decomposition of ethanol

$$C_2H_5OH(g) \xrightarrow{Cu} CH_3CHO(g) + H_2(g)$$

$$C_2H_5OH(g) \xrightarrow{Al_2O_3} C_2H_4(g) + H_2O(g)$$

2 Decomposition of methanoic acid

$$HCOOH(g) \xrightarrow{Al_2O_3} CO(g) + H_2O(g)$$

$$HCOOH(g) \xrightarrow{ZnO} CO_2(g) + H_2(g)$$

Autocatalysis
Autocatalysed reactions are reactions in which one of the products formed acts as a catalyst. For example the oxidation of ethanedioate ions by acidified manganate(VII) ions is catalysed by the manganese(II) ions produced during the reaction.

$$2MnO_4^- + 16H^+ + 5C_2O_4^{2-} \longrightarrow 2Mn^{2+} + 8H_2O + 10CO_2$$

In reactions of this type the graph obtained on plotting rate against time has a maximum.

Enzymes
Most biochemical reactions are controlled by catalysts called enzymes. These are often proteins. Since their structure is readily broken down on heating, Arrhenius' equation is not obeyed.

(e) Radiation

The rate of certain reactions is affected by the intensity and wavelength of the incident radiation. The combination of hydrogen and chlorine to form hydrogen chloride is explosive in ultra-violet light but more controlled in diffuse light. Nernst (1916) suggested that many of these reactions proceed by a chain mechanism involving free radicals. A free radical is an atom or group of atoms containing an odd number of electrons. It is uncharged and very reactive.

Example

Hydrogen, chlorine reaction

Initiation	Cl_2	$\xrightarrow[h\nu]{\text{quantum of light}}$	$2Cl\cdot$
Propagation	$Cl\cdot + H_2$	$\longrightarrow$	$HCl + H\cdot$
	$H\cdot + Cl_2$	$\longrightarrow$	$HCl + Cl\cdot$ etc.
Termination	$H\cdot + H\cdot$	$\longrightarrow$	H_2
	$Cl\cdot + Cl\cdot$	$\longrightarrow$	Cl_2
	$Cl\cdot + H\cdot$	$\longrightarrow$	HCl

The quantum efficiency of the reaction is the number of molecules decomposing per quantum of light. For this reaction the quantum efficiency is very high having a value of between 10^4 and 10^6. In contrast the quantum efficiency for the formation of hydrogen bromide from hydrogen and bromine is much lower, being only 0·02.

Initiation	Br_2	$\xrightarrow{h\nu}$	$2Br\cdot$	①
Propagation	$Br\cdot + H_2$	$\longrightarrow$	$HBr + H\cdot$	②
	$H\cdot + Br_2$	$\longrightarrow$	$HBr + Br\cdot$	③
Termination	$Br\cdot + Br\cdot$	$\longrightarrow$	Br_2	④
	$H\cdot + H\cdot$	$\longrightarrow$	H_2	⑤
	$H\cdot + Br\cdot$	$\longrightarrow$	HBr	⑥

Reaction ② is endothermic and very slow. Thus reaction ④ is the more probable reaction.

Theories of reaction kinetics

1 **Collision theory** (see page 165).

2 **Transition state theory**

This theory suggests that an intermediate known as the activated complex or transition state is formed during the reaction. This complex cannot be isolated and represents the state of the reacting molecules at the peak of the energy profile curve.

e.g. (a) Hydrolysis of bromoalkanes

(i) 1-Bromobutane

The hydrolysis of 1-bromobutane is found to be first order with respect to both OH^- ions and 1-bromobutane.

$$CH_3CH_2CH_2CH_2Br + OH^- \longrightarrow CH_3CH_2CH_2CH_2OH + Br^-$$

This may be explained by assuming that in the transition state the incoming OH^- ion and outgoing Br^- ion are equally associated with the hydrocarbon group. The energy profile for the reaction is shown below.

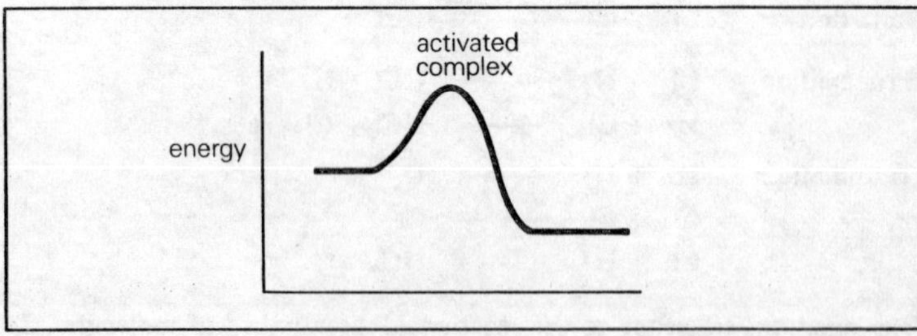

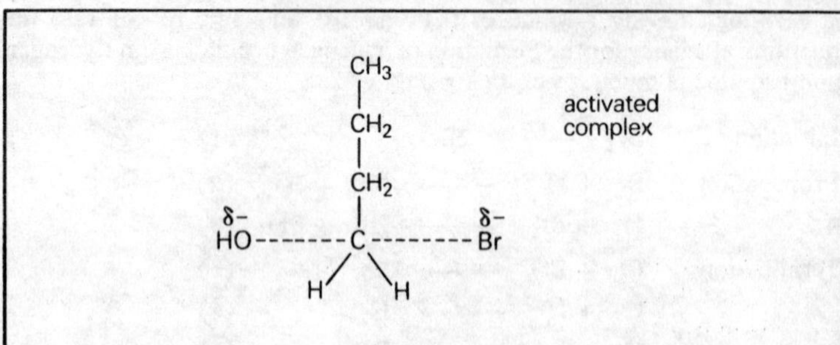

(ii) 2-bromo-2-methylpropane

The alkaline hydrolysis of 2-bromo-2-methyl propane is found to be zero order with respect to OH^- ions but first order with respect to 2-bromo-2-methylpropane.

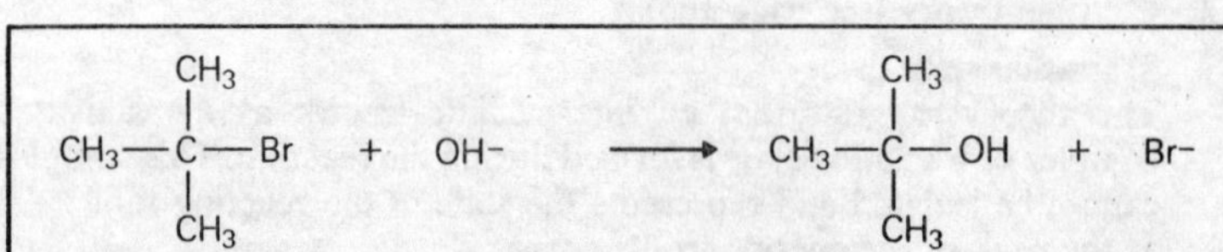

This is explained by assuming that there are 2 stages in the reaction.
The first is slow and rate-determining and involves the formation of a carbonium ion.

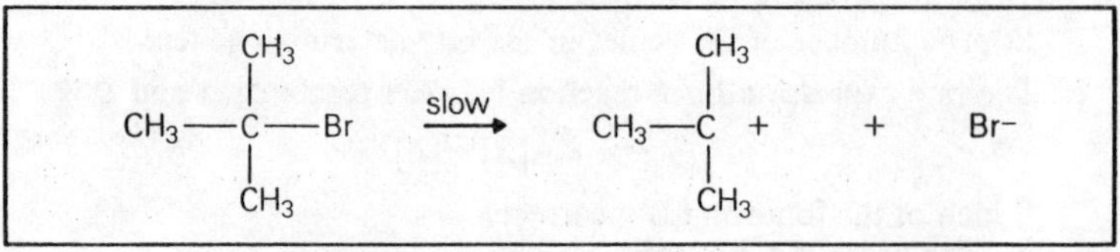

The second is fast and involves the addition of an OH^- ion to the carbonium ion.

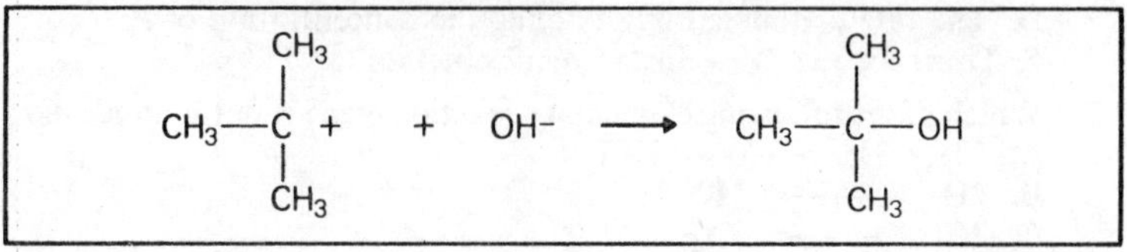

A diagram of the energy profile for the reaction is shown below.

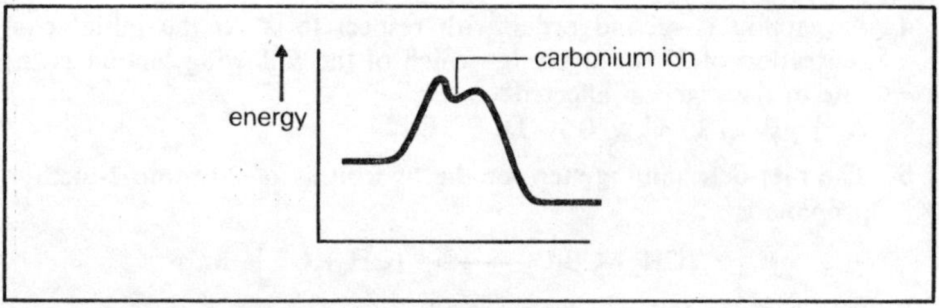

(b) The decomposition of hydrogen iodide
It has been assumed that the reaction occurs via the activated complex shown.

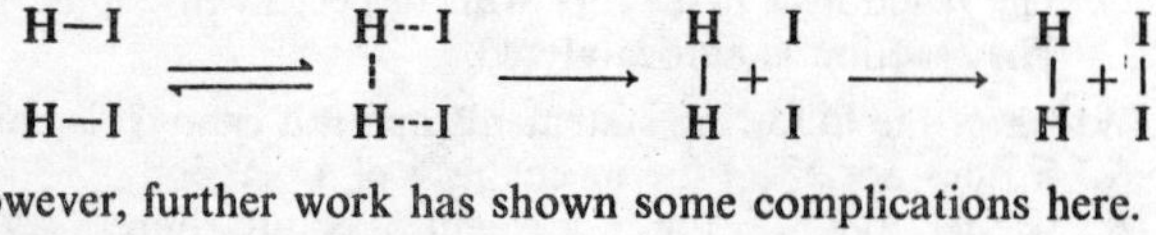

However, further work has shown some complications here.

Practice questions

In questions 1 to 7 give the letter of the correct answer.

1 Which of the following is a correct definition of the overall order of a reaction?

A. The number of molecules in the fastest step.

B. The molecularity of the initial step.
C. The sum of the powers of the concentration terms in the rate equation.
D. The ratio of the number of molecules of reactants to the number of molecules of products.
E. The number of molecules in the rate-determining step.

2 The rate expression for a reaction between reactants A and B is

$$\text{rate} = k[A]^{1/2}[B].$$

Which of the following is incorrect?
A. The order is 1·5 overall.
B. The order is 0·5 with respect to A.
C. The rate is halved by halving the concentration of A.
D. The rate is doubled by doubling the concentration of B.
E. Units for the rate constant could be $\text{mol}^{-1/2}\ \text{l}^{1/2}\ \text{s}^{-1}$.

3 Which of the following elementary reaction steps is not bimolecular?
A. $Cl\cdot + H\cdot \longrightarrow HCl$
B. $2H\cdot \longrightarrow H_2$
C. $2Cl\cdot \longrightarrow Cl_2$
D. $H^+ + HCOO^- \longrightarrow HCOOH$
E. $N_2O_4 \longrightarrow 2NO_2$

4 A reaction is second order with respect to X. If the initial concentration of X is halved, by which of the following factors is the rate of the reaction affected?
A. $\frac{1}{4}$ B. $\sqrt{2}$ C. $\sqrt{0{\cdot}5}$ D. $\frac{1}{2}$ E. 2

5 The rate-determining step for the hydrolysis of 2-bromo-2-methylpropane is

$$(CH_3)_3CBr \longrightarrow (CH_3)_3C^+ + Br^-.$$

Which of the following is correct?
A. Increasing concentration of $(CH_3)_3CBr$ decreases the rate.
B. The reaction is uni-molecular.
C. The reaction is first order with respect to OH^- ions.
D. The reaction is first order with respect to Br^- ions.
E. The reaction is autocatalysed.

6 Which of the following statements about a catalyst is correct?
A. It does not affect the mechanism of a reaction.
B. It alters the position of a chemical equilibrium.
C. It remains unchanged physically at the end of the reaction.
D. It changes the overall activation energy of a chemical reaction.
E. It alters the order and molecularity of a chemical reaction.

7 Which of the following statements about the photochemical reaction between gaseous hydrogen and chlorine is false?
A. The reaction proceeds by a free-radical mechanism.
B. The reaction may be initiated by blue but not by red light.

C. The reaction is slower in diffused than in ultraviolet light.
D. Photons cause the dissociation of hydrogen molecules.
E. Termination may be caused by the combination of two chlorine radicals.

Questions 8 to 10
In questions 8 to 10 give the letter of the graph A to E which corresponds to the reaction in the question

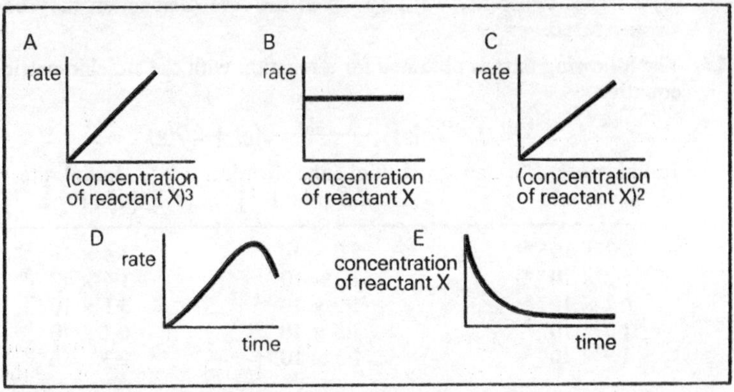

8 $^{238}_{92}U \longrightarrow \, ^{234}_{90}Th + \, ^{4}_{2}He$

9 $2MnO_4^- + 16H^+ + 5C_2O_4^{2-} \longrightarrow 2Mn^{2+} + 8H_2O + 10CO_2$

10 A zero order reaction.

11 Draw a diagram to represent the energy profile of the reversible reaction

$$A(g) + B(g) \rightleftharpoons C(g) + D(g)$$

The forward reaction is exothermic.
(a) On the diagram, show (i) the activation energy of the forward and backward reactions, (ii) ΔH for the forward reaction.
(b) Give two ways, with explanations, of how the initial rate of the forward reaction could be increased without changing the initial concentrations of A and B.
(c) Give two ways in which the equilibrium may be displaced to the right.

12 The overall reaction $2W + X \rightarrow Y + Z$ occurs by 2 elementary steps via an intermediate species A.

$$W + X \longrightarrow A \quad (1)$$

$$A + W \longrightarrow Y + Z \quad (2)$$

The rate of the overall reaction is given by

$$\text{rate} = k[W]^2[X]$$

(a) Define 'molecularity' and give the molecularity of steps ① and ②.
(b) Define 'order of a reaction' and give the overall order of the reaction.
(c) If the concentration of only 1 of the reactants is doubled give, with reasons, which would give the greater increase to the overall rate.
(d) Discuss two other ways by which the rate of reaction may be increased.

13 The following data is obtained for a reaction with the stoichiometric equation

$$2A(g) + 2B(g) \longrightarrow C(g) + 2D(g)$$

Initial concentration of A (mol l^{-1})	Initial concentration of B (mol l^{-1})	Initial rate (mol l^{-1} h^{-1})
$6{\cdot}2 \times 10^{-4}$	$5{\cdot}0 \times 10^{-3}$	$3{\cdot}4 \times 10^{-10}$
$6{\cdot}2 \times 10^{-4}$	$1{\cdot}0 \times 10^{-2}$	$1{\cdot}4 \times 10^{-9}$
$6{\cdot}2 \times 10^{-4}$	$1{\cdot}5 \times 10^{-2}$	$3{\cdot}1 \times 10^{-9}$
$1{\cdot}2 \times 10^{-3}$	$1{\cdot}5 \times 10^{-2}$	$6{\cdot}2 \times 10^{-9}$
$1{\cdot}9 \times 10^{-3}$	$1{\cdot}5 \times 10^{-2}$	$9{\cdot}3 \times 10^{-9}$

(a) Give the order of the reaction with respect to (i) A, (ii) B.
(b) Write the experimental rate law and calculate the specific rate constant for the reaction.
(c) Suggest a possible mechanism for the reaction.

14 The rate of the reaction $2A + B \rightarrow C + D$ is given by the equation

$$\text{rate} = k[A]^2[B]$$

where k is the rate constant.

State
(a) the units of k if the concentrations are in mol l^{-1},
(b) the effect on the rate when the concentration of A and B are simultaneously doubled.

Give a graphical representation of the distribution of velocities of gas molecules at different temperatures and use it to explain the effect of temperature on reaction rate.
Distinguish between homogeneous and heterogeneous catalysis giving an example of each.
Give an example in which a specific catalyst may alter the route of a reaction.

11 Answers

Chapter 1

1 A **2** C **3** E **4** E **5** A **6** D **7** B
8 B **9** B **10** E **11** B **12** A **13** C **14** E
15 D **16** $W = {}^{223}_{88}\text{Ra}$, $X = {}^{14}_{6}\text{C}$, $Y = {}^{6}_{3}\text{Li}$, $Z = {}^{228}_{89}\text{Ac}$.

Chapter 2

1 D **2** A **3** C **4** C **5** A **6** B **7** A
8 E **9** D **10** C

Chapter 3

1 A **2** B **3** D **4** E **5** C **6** C **7** D
8 E **9** A **10** B **11** E **12** D **13** C
14 $\Delta H_c^\ominus = -1368{\cdot}1\ \text{kJ mol}^{-1}$

Chapter 4

1 E **2** A **3** D **4** D **5** A **6** B **7** C
8 D **9** D **10** E **11** A **12** A **13** B **14** B
15 C **16** E **17** A **18** C

Chapter 5

1 D **2** A **3** E **4** A **5** D **6** D **7** E
8 E **9** A **10** C

Chapter 6

1 E **2** C **3** C **4** B **5** B **6** E **7** A
8 A **9** D **10** A **11** D **12** (b) 76·9% 10·3

Chapter 7

1 D **2** E **3** D **4** B **5** E **6** C **7** D
8 E **9** D **10** D

Chapter 8

1 E **2** D **3** D **4** B **5** C **6** A **7** E
8 D **9** A **10** B **11** C **12** B **13** B **14** A
15 (a) (i) 4 (ii) 1·52 (iii) 4·28,
(b) (i) 10^{-8} 10^{-6} (ii) 0·63 10^{-14} (iii) 5×10^{-13} 2×10^{-2},
(c) (i) 13·3 (ii) 12·6 (iii) 0 (iv) 1·30 (v) 2·88 (vi) 6·39
16 (c) $1{\cdot}66 \times 10^{-4}$, (d) 3·79

Chapter 9

2 (i) $G = 89$ kJ (ii) $s = -108$ J K^{-1} mol l^{-1}
3 $p_{CO_2} = 1{\cdot}0 \times 10^{-23}$ atm **4** $K_c = 3{\cdot}4$ l mol^{-1}
5 12·14 J K^{-1} mol l^{-1}
7 (a) (i) +2·32 (ii) $2{\cdot}09 \times 10^2$ atm, (b) $1{\cdot}23 \times 10^3$ atm, (c) 1500 K

Chapter 10

1 C **2** C **3** E **4** A **5** B **6** D **7** D
8 E **9** D **10** B

NOTES

EXAM RESULTS
TH
SION AIDS

Literature Revision Notes and Examples
£1.50 each
Titles: *Chaucer's Prologue, Merchant of Venice, Julius Caesar, Richard II, A Midsummer Night's Dream.*

New Testament Studies
£1.50 each
Titles: *St. Matthew, St. Mark, St. Luke, St. John, Acts of the Apostles.*

Law Revision Notes
£1.95
Titles: *Principles of Law, Criminal Law, Family Law, Law of Tort, Company Law.*

Model Answers
£1.25 each
Titles: *Julius Caesar, Macbeth, Romeo & Juliet, Merchant of Venice, Aids to Mathematics, Aids to Essay Writing, English Language, Model Essays, Practice in Summary Writing, Essay Plans, Biology, Chemistry, Human Biology, Physics, Mathematics, Commerce, Economics, British Isles Geography, British Economic History, Accounts, Commercial Mathematics, Integrated Science.*

Test Yourself
95p each
Titles: *English Language 1, English Language 2, French, German, Commerce, Economics, Chemistry, Physics, Mathematics, Modern Mathematics, Biology, Human Biology, St. Matthew, St. Mark, St. Luke, St. John, Acts of the Apostles, Commercial Mathematics, Accounts, Statistics, British Isles Geography, British Economic History.*